Lecture Notes in Mathematics

History of Mathematics Subseries

Volume 2392

Providing captivating insights into facets of the recent history of mathematics, the volumes in this subseries of LNM explore interesting developments of the past 200 years or so of research in this science. Their aim is to emphasize the evolution of its intellectual discourse, the emergence of new concepts and problems, the ground-breaking innovations, the human interactions, and the surrounding events that all contributed to weave the backdrop of today's research and teaching in mathematics. These high-level and largely informal accounts will be of interest to researchers and graduate students in the mathematical sciences, in the history or the philosophy of mathematics, and to anyone who seeks to understand the historical growth of the discipline.

Ian G. Macdonald · Anne-Marie Aubert

Spherical Functions on a Group of *p*-adic Type

Second Edition

Ian G. Macdonald
Steventon, UK

Anne-Marie Aubert
Institut de Mathématiques de Jussieu-Paris Rive Gauche
Sorbonne Université and Université Paris Cité, CNRS
Paris, France

ISSN 0075-8434 ISSN 1617-9692 (electronic)
Lecture Notes in Mathematics
ISSN 2193-1771 ISSN 2625-7157 (electronic)
History of Mathematics Subseries
ISBN 978-3-032-15670-9 ISBN 978-3-032-15671-6 (eBook)
https://doi.org/10.1007/978-3-032-15671-6

Mathematics Subject Classification: 20C08, 20E42, 22E50

This Springer imprint is published by the registered company Springer Nature Switzerland AG
The registered company address is: Gewerbestrasse 11, 6330 Cham, Switzerland

Foreword to the New Edition

Ian Grant Macdonald (11 October 1928 – 8 August 2023) wrote his book *Spherical Functions on a Group of p-adic Type* (1971) following his lecture course on the subject at the Ramanujan Institute in Madras in the spring of 1970. In the text, Macdonald extends the results of his article "Spherical functions on a p-adic Chevalley group" (*Bulletin of the A.M.S* **74** (1968), 520–525) to a general theory of zonal spherical functions on the group of rational points of a simply-connected simple algebraic group defined over a p-adic field, and relative to any maximal compact subgroup. Spherical functions are very classical objects of study, with deep roots in the "prehistory" of harmonic analysis, the theory of spherical harmonics. Classical special functions such as Legendre polynomials – a special type of hypergeometric function – had been studied in this context for a long time, with a multitude of important applications in physics. Despite these very classical origins, Macdonald's book on spherical functions was highly relevant and timely. The relevance of spherical functions in the modern theory of harmonic analysis on reductive groups had been pointed out by mathematicians such as I.M. Gel'fand, R. Godement, Harish-Chandra and I. Satake. Langlands' emerging theory of automorphic L-functions had moved the subject even further to the center of the stage, while at the same time pointing out the crucial role of the p-adic case.

Macdonald's book is a classic gem. It is a prime example of the wonderful qualities of Ian Macdonald as a mathematician. In a period when the subject was gaining in prominence, Macdonald was able to bring the many ideas that were floating around together, and write the definitive treatise. The text is concise and efficient, yet exquisitely transparent and lucid, and is packed with new insights. Macdonald chose to treat the required structure theory of semisimple groups defined over a non-Archimedean local field axiomatically, as a group with an affine BN pair, based on the theory developed by Bruhat and Tits. His elegant discussion of the structure theory of the group and the theory of the building is a delight in its own right, making the text accessible for readers with only a basic knowledge of reductive groups. It is an excellent introduction to the theory of Bruhat and Tits, matching it with the deep application towards zonal spherical functions at the same time.

In his work towards the Plancherel formula of real reductive groups, Harish-Chandra had encountered several technical difficulties due to the fact that the zonal spherical functions in the real case are usually transcendental functions, which can only be described indirectly. The all important c-function which Harish-Chandra had introduced in this context, crucial for expressing the spherical Plancherel measure, was not available in explicit form at the time, impeding Harish-Chandra to prove the required estimates. This was resolved in the real case when Gindikin and Karpelevic proved their famous product formula in 1962, expressing Harish-Chandra's c-function as a product over all positive roots of certain beta-functions. In his book on the p-adic case, Macdonald revealed several surprising facts. He found an explicit formula for the zonal spherical function, showing that it is an elementary function in all cases now. His formula shows that the leading asymptotic term of the spherical functions at infinity is *equal* to the spherical function itself, in stark difference with the real case. The explicit formula can also be interpreted as an explicit inversion formula for the Satake transform.

Macdonald was a master at grasping the hidden meaning in explicit formulas. In the text, he points out that his formula for the Harish-Chandra c-function in the p-adic case and the resulting formula for the spherical Plancherel measure, can be expressed in terms of local zeta functions. He observes that this yields a uniform expression for the spherical Plancherel measure when his formula is compared with that of Gindikin and Karpelevic. This remark is undoubtedly indicative of Macdonald's sense of the deep connections in mathematics, and how explicit formulas sometimes reveal these connections. This specific theme has played an important role in his development of the theory of what is now known as *Macdonald polynomials*, a unified theory in which both the p-adic and the real case exist as special limit cases. His ability to recognise deep connections in mathematics through explicit formulas has led him to several inspired conjectures, which have served as important guidance in the modern developments of representation theory, algebraic geometry, harmonic analysis and modular forms.

I am quite excited about the beautifully annotated new edition of the original text, which has been reworked with great dedication by Anne-Marie Aubert. She has preserved to a large extend the structure of the main body of the original text. Most importantly, her annotations form a treasure trove of connections of Macdonald's classical text to modern topics in representation theory. I sincerely hope that this new edition of Macdonald's text will serve as a source of inspiration and an invitation to this beautiful subject.

Korteweg-de Vries Instituut, *Eric Opdam*
Universiteit van Amsterdam

Preface to the New Edition

This new edition contains the original notes by Ian Grant Macdonald (originally published by the University of Madras in November 1971), to which I have made the following small changes:

- In the original edition, notes for each chapter were collected together in a single section at the end of the main text. These notes have now been relocated to their respective chapters, appearing as a new section titled "Notes on Chapter n" ($n = \mathrm{I}, \ldots, \mathrm{V}$).
- Axiom V has been corrected along with a remark that was made by Pierre Deligne to Ian G. Macdonald (see footnote 8 in §2.5).
- I chose to denote by Λ (instead of T) the group of translations of an affine Weyl group, and by G_F (instead of P_F) the parahoric subgroup associated to the facet F.
- I turned the notation Q into the more standard notation P_{W_0} for the Poincaré polynomial of the Coxeter group W_0.

I also added the following complements:

- An extended table of contents.
- An introductory chapter which explains the importance and historical context of these notes.
- A new section "Additions to Chapter n" to each to original Chapter n ($n = \mathrm{I}, \ldots, \mathrm{V}$), mentioning applications of the notes in the literature and further developments of the subject.
- Some footnotes (all footnotes are new).
- An additional bibliography.
- An index of notation and terminology.

I would like to warmly thank Eric Opdam for the text he has written in honor of Ian G. Macdonald and for several helpful comments, and Yongqi Feng for his precious help in the preparation of these notes.

Paris, 3 October 2025 *Anne-Marie Aubert*

Preface to the First Edition

The present volume is based on a course of lectures given (by Ian Macdonald) at the Ramanujan Institute of Advanced Study in Mathematics at Madras in March–April 1970. In essence, it is an account of the theory of zonal spherical functions on the group of rational points of a simply-connected simple algebraic group defined over a non-Archimedean local field, relative to a suitably chosen maximal compact subgroup. However, partly in order to reduce to a minimum the mathematical prerequisites for the course, and partly in order to emphasize the formal structure of the theory, we have presented it in an axiomatic form: it turns out that the axioms of Bruhat and Tits are peculiarly well adapted to our purposes, and we have therefore used them as a basis of our exposition. Consequently, the reader need not even know the definition of an algebraic group in order to understand what is going on.

I. G. Macdonald

Contents

Introduction

Sophus Lie initiated the study of continuous transformation groups. His work laid the groundwork for understanding symmetries and transformations in various mathematical and physical contexts. Continuous transformation groups evolved into what are now known as *Lie groups* and *Lie algebras*: Lie groups are groups with a smooth manifold structure, and their corresponding Lie algebras capture their infinitesimal structure.

Lie groups and algebraic groups are important in many major areas of mathematics and mathematical physics. They occur in particular as groups of automorphisms of geometric structures, as symmetries of differential systems, or as basic tools in the theory of automorphic forms.

A Lie group, by definition, is a topological group that is also a smooth manifold, and a manifold is locally homeomorphic to a Euclidean space (that is, around each point, there exists a neighborhood which is homeomorphic to an open set in Euclidean space). Since Euclidean spaces are locally compact, and homeomorphisms (that is, continuous bijections with continuous inverses) preserve local compactness, Lie groups are locally compact. (A topological space is locally compact if every point has a neighborhood with a compact closure.)

Another important source of locally compact topological groups is provided by the groups known as *p-adic (reductive) groups*. These are the groups called by Macdonald "groups of p-adic type". Here p is a fixed prime number. The p-adic numbers form a topological field, denoted as $\mathbb{Q}_p$, that is locally compact. A nondiscrete locally compact topological field is said to be a *local* field. The fields $\mathbb{R}$, $\mathbb{C}$ are called the *Archimedean* local fields, while the finite extensions of $\mathbb{Q}_p$ are called *non-Archimedean*. The field $\mathbb{Q}_p$ (and its extensions of finite degree) has characteristic zero. With the extensions of finite degree of the field $\mathbb{F}_p((t))$, with t an indeterminate, they constitute all the non-Archimedean local fields.

A p-adic group is the group of k-rational points of a connected reductive algebraic group defined over a non-Archimedean local field k. Examples of p-adic groups are the general linear group $\mathrm{GL}_n(k)$, where n is a positive integer, defined to be the group of $n \times n$-matrices with entries in the field k and non-zero determinant, the special linear group $\mathrm{SL}_n(k) \subset \mathrm{GL}_n(k)$ of matrices with determinant equal to 1, and the projective linear group $\mathrm{PGL}_n(k)$, the quotient of $\mathrm{GL}_n(k)$ by its centre $k^\times$.

Let $\overline{k}$ be an algebraically closed field, and let $\mathbf{G}$ be a connected reductive algebraic group over $\overline{k}$. Let $\mathbf{B}$ be a Borel subgroup of $\mathbf{G}$, let $\mathbf{T}$ be a maximal torus of $\mathbf{B}$ and let $\mathbf{N}$ be the normalizer of $\mathbf{T}$ in $\mathbf{G}$. The quotient $W_0 := \mathbf{N}/\mathbf{T}$ is a finite *Weyl group* (named after Hermann Weyl). For instance, if $\mathbf{G}$ is $\mathrm{GL}_n(\overline{k})$ or $\mathrm{SL}_n(\overline{k})$, the group W_0 is isomorphic to the symmetric group $\mathfrak{S}_n$.

The Weyl group of a *root system* is a subgroup of the isometry group of that root system: it is the subgroup which is generated by reflections through the hyperplanes orthogonal to at least one of the roots, and as such is a finite reflection group. Abstractly, Weyl groups are Coxeter groups, and are important examples of them.

The group $\mathbf{G}$ is the disjoint union $\mathbf{G} = \bigsqcup_{w \in W_0} \mathbf{B}\dot{w}\mathbf{B}$, where $\dot{w}$ is a representative of w in $\mathbf{N}$. This result, known as the *Bruhat decomposition* of $\mathbf{G}$, is a restatement of Theorem 7.1 in Bruhat's 1956 article [Bru1] (in which k was assumed to be equal to $\mathbb{C}$ and $\mathbf{G}$ to be semisimple). It gives rise to the decomposition of the flag variety $\mathbf{G}/\mathbf{B}$ into *Schubert cells*, and allows many questions about $\mathbf{G}$ to be reduced to questions about the Weyl group W_0.

The Bruhat decomposition was discovered rather late in the history of Lie groups. It was preceded by Ehresmann's discovery [Ehr] of a closely related cell decomposition for flag manifolds, and was axiomatized by Tits in the notion of a *group with a (B, N)-pair* or *Tits system*. A precise account of the history of can be found in [Lus5].

The notion of (B, N)-pair is an extremely important generalization of that of a Coxeter group, and indeed every (B, N)-pair gives rise to a Coxeter group. Coxeter groups always act on simplicial complexes whose geometry is closely connected with their properties. Similarly, a group with a (B, N)-pair also acts on a simplicial complex, the *Tits building* or also (in the p-adic case) the *Bruhat–Tits building*. These fundamental notions are introduced and studied in Chapter II of the book, the additional section of which describes several important applications.

Apart from strong topological differences, real and non-Archimedean simple Lie groups share many combinatorial and geometric properties. From the geometric point of view they both act, with strong transitivity properties, on contractible spaces carrying nice non-positively curved complete distances. In the real case, these are the symmetric spaces. The corresponding spaces for groups over totally disconnected local fields are Bruhat–Tits buildings. They are unions of Euclidean tilings, called apartments, which are the analogues of maximal flats in symmetric spaces. The Bruhat–Tits building is a geometric object that plays a central role in the structure theory of reductive groups over non-Archimedean fields.

In general, a "building" is a gluing of (poly)simplicial subcomplexes, all isomorphic to a given tiling naturally acted upon by a Coxeter group. The copies of this tiling in the building are called apartments and must satisfy, by definition, strong incidence properties which make the whole space very symmetric.

Affine buildings first emerged from work of Iwahori and Matsumoto [IM] for Chevalley groups over p-adic fields. The general theory was then developed by Bruhat and Tits [BT1], and a classification was finally given by Tits in [T3]. One of the main results obtained in [BT1] is that the group $\mathbf{G}(k)$ of rational points of a connected reductive algebraic group $\mathbf{G}$ over a suitable valued field k possesses valued root data (and hence the desired building $\mathcal{B}(\mathbf{G}, k)$ on which the group G acts naturally). Key ingredients are affine group schemes over the ring of integers of k, having G as generic fiber and having as group of integral points the stabilizer of a suitable subset of the Bruhat–Tits building $\mathcal{B}(\mathbf{G}, k)$ of G. When k is a non-Archimedean local field, $\mathcal{B}(\mathbf{G}, k)$ is *Euclidean*, meaning that its apartments are Euclidean tilings. As a consequence, $\mathcal{B}(\mathbf{G}, k)$ carries a natural non-positively curved metric, which, when G is semisimple, allows one to classify in a geometric way its maximal bounded subgroups.

Bruhat–Tits buildings also have several applications in group theory that are of geometric or topological interest. The simplest buildings are trees, and can be used to prove Ihara's Theorem that every torsion-free discrete subgroup of $\mathrm{SL}_2(\mathbb{Q}_p)$ is free (see [Ih]). They have applications in many other fields, notably in the theory of automorphic forms, where local zeta functions of Shimura varieties are related to properties of associated Bruhat–Tits buildings, for cohomology computations in the series of papers by Raghunathan and G. Prasad, in questions on the classification of bundles by Raghunathan and Ramanathan, and in the work of Garland and Raghunathan on loop spaces.

In a different direction, random walks on Bruhat–Tits buildings have been studied for over thirty years, see for instance [Par3, AST], and the description obtained in [Tr] of the asymptotic behavior of the Green function of a finite range isotropic random walk. Bruhat–Tits buildings also occur in noncommutative geometry via the Baum–Connes assembly map, as a special case of classifying spaces of proper actions (see [AJV, Laf]).

The theory of spherical functions, particularly *zonal spherical functions*, developed as a part of the broader theory of harmonic analysis on Lie groups, providing tools for analyzing functions and distributions on these groups. Zonal spherical functions on p-adic groups play a foundational role in the harmonic analysis and representation theory of these groups. Their importance is comparable to that of spherical harmonics in the real Lie groups setting.

Spherical functions associated with compact semisimple Lie groups were introduced and studied with success in the now classical works of E. Cartan and Weyl. Early examples of their analogues for such non-compact groups appeared in the works of Bargmann [Bar] and of Gelfand and Naimark [GeNa]. In a general setting they were first introduced by Gelfand [Gel] and can be described roughly as functions on a (not necessarily compact) group which have simple transformation properties with respect to a given (usually compact) subgroup. These functions were then extensively studied by Harish-Chandra [HC, 8, 9], Godement [7] and others.

In the case of a real reductive group G with complexified Lie algebra $\mathfrak{g}$, zonal spherical functions are defined as eigenfunctions for the action of the center $Z(\mathfrak{g})$ of the universal enveloping algebra $U(\mathfrak{g})$ of $\mathfrak{g}$ on the smooth functions on G that are constant on double K-classes, with K a *maximal compact subgroup* of G, that is, K is a compact subgroup that is not properly contained in any other compact subgroup of G. While not always unique, maximal compact subgroups of G are unique up to conjugation by an element of G.

The representation theory of semisimple Lie algebras and Lie groups has been closely intertwined with the theory of symmetric spaces since E. Cartan's pioneering work in the 1920s. A beautiful classical result shows that the zonal spherical functions of these spaces can be identified with a family of orthogonal polynomials. With the introduction of quantum groups in the 1980s, it was natural to look for and study quantum versions of symmetric spaces. This search became especially compelling as q-orthogonal polynomials, which are obvious candidates for quantum zonal spherical functions, began to appear in the literature (see for example [Mac2, Ko]).

Maximal compact subgroups of p-adic groups are defined similarly as for real reductive groups. However, maximal compact subgroups of a p-adic group G are not always unique up to conjugation by an element of G, e.g., for k a non-Archimedean local field, the maximal compact subgroups of $\mathrm{SL}_n(k)$ are conjugated under $\mathrm{GL}_n(k)$ but not under $\mathrm{SL}_n(k)$. Also, in the context of p-adic groups, compactness is closely related to the concept of "openness": a maximal compact subgroup is always an open subgroup of the larger p-adic group. Bruhat summarized in [Bru5] several results on maximal compact subgroups in a p-adic semisimple group and raised the following three fundamental questions: (a) existence; (b) classification up to inner automorphisms; (c) existence of at least one "good" class (having certain favorable properties such as the analogue of the Iwasawa decomposition and the elementary divisor decomposition). He gave in [Bru3] a solution of (c) under the completeness assumption for the case of Chevalley groups.

Spherical functions on p-adic groups are functions that satisfy certain invariance properties under maximal compact subgroups. They provide a framework for analyzing functions on the p-adic group, analogous to spherical harmonics in Euclidean spaces, and are closely related to the study of representations of the group. They were first studied in 1958 by Mautner, in [Mau2, Mau3], who established among other things the inversion formula and the Plancherel formula of the Fourier transform for a special class of spherical functions on the group $\mathrm{PGL}_2(\mathbb{Q}_p)$. This was the beginning of the representation theory of p-adic reductive groups. Satake showed in [14] that the principal part of the theory holds for a large class of reductive algebraic groups over p-adic fields, containing all simple classical groups without center. I.G. Macdonald did foundational work on spherical functions on p-adic Chevalley groups. He found analogues of many of Harish-Chandra's results for p-adic groups, including explicit formulas for spherical functions and the associated "c-functions" in that setting. He showed that the spherical Plancherel measure is given in terms of c-functions, and provided explicit product formulas in many unramified cases, which are analogous to Gindikin–Karpelevich type formulas. Their explicit forms are useful in number theory, particularly in the computation of automorphic L-functions.

In the representation theory of reductive p-adic groups the Hecke algebra $\mathcal{H}(G)$ of compactly supported smooth functions on a p-adic group often plays a role similar to that of the universal enveloping algebra $U(\mathfrak{g})$, and the *Bernstein center* plays the role of $Z(\mathfrak{g})$ in the representation theory of real reductive groups.

Zonal spherical functions relative to a maximal compact subgroup K of G arise as matrix coefficients of K-invariant vectors in irreducible representations of G (see [Car]). They encode the action of the Hecke algebra on the space of K-invariant vectors in these representations. Spherical functions are essential in the determination of the local Langlands correspondence for unramified representations.

The *K-spherical Hecke algebra* of G is the subalgebra $\mathcal{H}(G, K)$[1] of $\mathcal{H}(G)$ consisting of the compactly supported locally constant K-biinvariant functions on G with complex values. For $G = \mathrm{GL}_2(\mathbb{Q}_p)$ and $K = \mathrm{GL}_2(\mathbb{Z}_p)$, where $\mathbb{Z}_p$ is the ring of integers of $\mathbb{Q}_p$, the algebra $\mathcal{H}(G, K)$ is essentially the portion corresponding to the prime p of the original Hecke algebra introduced by Hecke in [He] to explain the appearance of Euler products as the L-series of automorphic forms. The *Satake isomorphism* identifies the algebra $\mathcal{H}(G, K)$ with a polynomial ring related to the *Langlands dual group* of G (see Theorem 3.3.6). Zonal spherical functions diagonalize $\mathcal{H}(G, K)$, providing eigenfunctions with eigenvalues corresponding to Langlands parameters of spherical representations of G.

In the representation theory of reductive p-adic groups the Hecke algebra $\mathcal{H}(G)$ of compactly supported smooth functions on a p-adic group G often plays a role similar to that of the universal enveloping algebra $U(\mathfrak{g})$, and the *Bernstein center* plays the role of $Z(\mathfrak{g})$ in the representation theory of real reductive groups.

There are strong connections with Lie combinatorics associated to parametrization of the representations of real Lie groups and p-adic groups. This phenomenon is illustrated for instance by the case of the Fourier transform on p-adic reductive groups. The starting point in spherical harmonic analysis was to exhibit a suitable Gelfand pair: this was done by Satake in [14], who also obtained a combinatorial parametrization of the spherical functions providing the desired Fourier transform [31]. The exact computation of the latter functions was achieved by Macdonald for Chevalley groups in 1967, leading to an explicit description of the involved Plancherel measure, in the first edition of this book (see Chapter V). The same formula, in a slightly more general situation, was obtained by Casselman in [Cas], as a consequence of general results on unramified representations of a p-adic reductive group G. The unramified representations form one of the most fundamental classes in the representation theory of G. Their importance lies in their relation to the theory of automorphic representations: almost all local components of an automorphic representation are unramified.

Zonal spherical functions relative to a maximal compact subgroup K of G arise as matrix coefficients of K-invariant vectors in irreducible representations of G (see [Car]). They encode the action of the algebra $\mathcal{H}(G)$ on the space of K-invariant

[1] The algebra $\mathcal{H}(G, K)$ is commutative, see Corollary 3.3.7.

vectors in these representations. Spherical functions are essential in the determination of the local Langlands correspondence for unramified representations.

The *c-function* appeared in Harish-Chandra's 1958 publication [8] on the asymptotics of zonal spherical functions on Riemannian symmetric spaces of non-compact type and its applications to the Plancherel formula on these spaces. This function plays a key role in understanding the Plancherel measure, a measure which is essential for Fourier analysis on Lie groups as it determines how functions are decomposed into their "spectral components". The Plancherel formula provides a precise way to decompose square-integrable functions on a group (or space) into irreducible representations, generalizing the classical Fourier transform. It is a central result in harmonic analysis, representation theory, and automorphic forms, especially over Lie groups and p-adic groups. In classical analysis, the Plancherel theorem says that the Fourier transform is an isometry on $L^2(\mathbb{R})$, that is, $\|f\|^2_{L^2(\mathbb{R})} = \|\hat{f}\|^2_{L^2(\mathbb{R})}$.

For a non-abelian group G, the Plancherel formula tells us how to express $L^2(G)$ in terms of irreducible unitary representations of G.

Just as for real semisimple Lie groups, one considers spherical principal series representations of a p-adic group. One looks at standard intertwining operators (between induced representations from opposite parabolic subgroups), and normalizes them via a spherical vector. The c-function (introduced and studied in §4.1) arises from the action of these intertwining operators on the spherical vector. It enters into: the spherical Plancherel formula (as the factor that describes how the measure behaves); expressions for constant terms of Eisenstein series; and product formulas (in particular, formulas akin to the Gindikin–Karpelevich formula) that express the c-function in terms of data coming from roots and the structure of the group.

The c-function plays a central role in the Plancherel formula for reductive groups over local fields (both p-adic and real), particularly in the context of spherical harmonic analysis. For a locally compact group G, the Plancherel formula gives an isometric isomorphism:

$$L^2(G) \simeq \int_{\hat{G}}^{\oplus} H_\pi \, \mathrm{d}\mu(\pi),$$

where $\hat{G}$ is the unitary dual of G (the set of equivalence classes of irreducible unitary representations), H_π is the Hilbert space of the representation π, the right-hand side is the direct integral of Hilbert spaces, and μ is the Plancherel measure.

Via what is called the Langlands–Shahidi method, Shahidi develops the theory of the local coefficient for irreducible generic representations of Levi subgroups of p-adic groups (see in particular [Sha1, Sha2]). The local coefficient is defined by means of the uniqueness property of Whittaker models paired with the theory of intertwining operators for representations obtained by parabolic induction from *generic* representations. It satisfies a functional equation and controls poles and zeros of intertwining operators. This yields analytic continuation, holomorphy, and product formulas, and plays a key role in describing the Plancherel measure. For unramified data, the local coefficient is closely related to the c-function. For ramified data, it is more elaborate, reflecting the deeper structure of the involved representation of G (including its conductor and ramification).

In more detail, let G be a unimodular separable locally compact topological group, and U a compact subgroup of G. A *zonal spherical function* (z.s.f. for short) on G relative to U is defined to be a complex-valued continuous function ω on G, which is bi-invariant with respect to U, maps the identity element of G to 1, and satisfies

$$f * \omega = \lambda_f \, \omega$$

for every complex-valued compactly-supported continuous function f on G, where $*$ is the convolution multiplication and λ_f a complex number depending on f.

A z.s.f. ω is said to be of *positive type* if

$$\int_G \int_G \omega(x^{-1}y)\phi(x)\overline{\phi(y)}\mathrm{d}x\mathrm{d}y$$

for all continuous functions $\phi\colon G \to \mathbb{C}$ with compact support. Let Ω (resp. Ω^+) denote the set of all z.s.f. (resp. z.s.f. of positive type) on G relative to U. By [6], there exists a unique positive measure $\mathrm{d}\omega$ on Ω^+, called the *Plancherel measure*, such that

$$\int_G |f(x)|^2 = \int_{\Omega^+} |\hat{f}(x)|^2 \mathrm{d}\omega$$

for every function f as above, and where $\hat{f}$ is the *Fourier transform of* f, i.e., the function. on Ω defined by

$$\hat{f}(g) := \int_G f(x)\omega(x^{-1})\mathrm{d}x\mathrm{d}y \geq 0.$$

Satake set up in [14] a general theory of zonal spherical functions on a reductive linear algebraic group over a p-adic field, and obtained integral formulas for zonal spherical functions.

When the book first appeared, it was one of a very small number of publications concerned with representations of p-adic groups. At just about that time, however, the subject began to be widely recognized as indispensable in understanding automorphic forms, and the literature on the subject started to grow rapidly. Numerous important questions in representation theory – among others, those connected with the Langlands–Shelstad fundamental lemma – are concerned with p-adic spherical functions (see notably [N], [CCH] and [Zha]).

Chapter I
Basic properties of spherical functions

1.1 The algebra $\mathcal{H}(G, K)$

Let G be a locally compact Hausdorff topological group with a countable basis of open sets. Let $C(G)$ denote the vector space of all continuous complex-valued functions on G, and let $C_c(G)$ be the subspace of $C(G)$ consisting of the functions with compact support.

Let K be a compact subgroup of G and let $C(G, K)$ be the subspace of $C(G)$ consisting of the functions f which are bi-invariant with respect to K: $f(gk) = f(kg) = f(g)$ for all $g \in G$ and $k \in K$. Also let

$$\mathcal{H}(G, K) = C(G, K) \cap C_c(G),$$

so that the functions in $\mathcal{H}(G, K)$ are continuous, compactly-supported and bi-invariant with respect to K.

Fix a left Haar measure $\mathrm{d}g$ on G, and define a multiplication (convolution) on $\mathcal{H}(G, K)$ as follows:

$$(f_1 * f_2)(g_0) = \int_G f_1(g_0 g) f_2(g^{-1}) \mathrm{d}g$$

for any $f_1, f_2 \in \mathcal{H}(G, K)$. In this way $\mathcal{H}(G, K)$ becomes an associative $\mathbb{C}$-algebra.

For the time being we shall not assume that $\mathcal{H}(G, K)$ is commutative, nor that G is unimodular.

Remarks

(i) If $f \in \mathcal{H}(G, K)$ and $\phi \in C(G, K)$ then $f * \phi \in C(G, K)$.

(ii) We shall be particularly concerned with the case where K is an *open* subgroup of G. In that case we shall always normalize the Haar measure on G such that $\int_K \mathrm{d}g = 1$.

I. G. Macdonald and A.-M. Aubert, *Spherical Functions on a Group of p-adic Type*, Lecture Notes in Mathematics 2392,
https://doi.org/10.1007/978-3-032-15671-6_1

Proposition 1.1.1 *Suppose that K is open in G. For each double coset $\xi = KgK$ let χ_ξ be the characteristic function of ξ. Then the χ_ξ form a basis of $\mathcal{H}(G, K)$ as a vector space over $\mathbb{C}$, and the algebra $\mathcal{H}(G, K)$ has an identity element, namely $\chi_K := \chi_1$.*

Proof Since K is open and compact, so is each double coset $\xi = KgK$, and therefore $\chi_\xi \in \mathcal{H}(G, K)$. Conversely, any $f \in \mathcal{H}(G, K)$ has compact support, which must therefore consist of a finite number of double cosets ξ, and hence f is a linear combination of the characteristic functions χ_ξ. Clearly these are linearly independent, which proves the first assertion.

Next, consider $\chi_K * \chi_\xi$. We have

$$(\chi_K * \chi_\xi)(g_0) = \int_G \chi_K(gg_0)\chi_\xi(g^{-1})\mathrm{d}g$$

and the integrand vanishes unless $g_0 g \in K$ and $g^{-1} \in \xi$, which implies that $g_0 \in \xi$. It follows that $\chi_K * \chi_\xi$ is a scalar multiple of χ_ξ, and one sees immediately that the scalar multiple is 1, by virtue of our normalization of the Haar measure (Remark (ii) above). Likewise $\chi_\xi * \chi_K = \chi_\xi$. □

Now let $\mathrm{d}k$ be a Haar measure on the compact group K, normalized so that $\int_K \mathrm{d}k = 1$. For any $\phi \in C_c(G)$ we define $\phi^\natural \in \mathcal{H}(G, K)$ by

$$\phi^\natural(g) = \int_K \int_K \phi(k_1 g k_2)\mathrm{d}k_1 \mathrm{d}k_2,$$

so that $\phi \mapsto \phi^\natural$ is a projection of $C_c(G)$ onto $\mathcal{H}(G, K)$.

Lemma 1.1.2 *Let $\omega \in C(G, K)$ be such that $\int_G g(g)\omega(g^{-1})\mathrm{d}g = 0$ for all $f \in \mathcal{H}(G, K)$. Then $\omega = 0$.*

Proof Let $\phi \in \mathcal{H}(G, K)$. Then

$$\begin{aligned}
\int_G \phi(g)\omega(g^{-1})\mathrm{d}g &= \int_G \phi(k_1 g k_2)\omega(g^{-1})\mathrm{d}g, \forall k_1, k_2 \in K), \\
&= \int_G \left(\int_K \int_K \phi(k_1 g k_2)\mathrm{d}k_1 \mathrm{d}k_2 \right) \omega(g^{-1})\mathrm{d}g \\
&= \int_G \phi^\natural(g)\omega(g^{-1})\mathrm{d}g \\
&= 0 \text{ (by hypothesis).}
\end{aligned}$$

Since ω is continuous, it follows that $\omega = 0$. □

1.2 Spherical measures and spherical functions

For each $f \in C_c(G)$ the norm

$$\|f\|_\infty = \sup_{g \in G} |f(g)|$$

is defined, since f is continuous and compactly supported. If C is any compact subset of G, let $\mathcal{H}_C(G)$ be the subspace of $C_c(G)$ consisting of the functions which vanish outside C. A (complex-valued) measure μ on G is by definition a linear functional $\mu\colon C_c(G) \to \mathbb{C}$ whose restriction to each $\mathcal{H}_C(G)$ is continuous with respect to the topology defined by the norm $\| \ \|_\infty$. Equivalently, for each compact subset C of G there exists a positive real number m_C such that

$$|\mu(f)| \leq m_C \|f\|_\infty \quad \text{for all } f \in \mathcal{H}_C(G).$$

As usual, we write $\int_G f(g)\mathrm{d}\mu(g)$ in place of $\mu(f)$, whenever convenient.

Definition A non-zero measure μ on G is said to be *spherical* (with respect to the compact subgroup K) if

(1) μ is bi-invariant with respect to K, in other words

$$\mathrm{d}\mu(kg) = \mathrm{d}\mu(gk) = \mathrm{d}\mu(g) \quad \text{for all } k \in K;$$

(2) $\mu\colon \mathcal{H}(G,K) \to \mathbb{C}$ is a ring homomorphism, i.e.

$$\mu(f_1 * f_2) = \mu(f_1)\mu(f_2) \quad \text{for all } f_1, f_2 \in \mathcal{H}(G,K).$$

For example, $f \mapsto \int_G f(g^{-1})\mathrm{d}g$ is a spherical measure.

Lemma 1.2.1 *If μ is spherical and $f \in C_c(G)$, then $\mu(f) = \mu(f^\natural)$.*

Indeed, this is true if μ is merely bi-invariant with respect to K. It shows that a spherical measure is uniquely determined by its restriction to $\mathcal{H}(G,K)$.

Proposition 1.2.2 *Suppose that K is open in G. Then every non-zero $\mathbb{C}$-algebra homomorphism $\nu\colon \mathcal{H}(G,K) \to \mathbb{C}$ determines a spherical measure on G.*

Proof For all $f \in C_c(G)$ define $\mu(f) = \nu(f^\natural)$. We have to check that μ satisfies the continuity condition above.

Let C be a compact subset of G. Since the double cosets KgK are open and mutually disjoint, C intersects only a finite number of them, say $Kg_iK (1 \leq i \leq n)$. Let χ_i be the characteristic function on Kg_iK. If $f \in C_c(G)$ has support contained in C, then $f^\natural$ vanishes outside $KCK = \bigcup_{i=1}^n Kg_iK$, and hence $f^\natural$ is a linear combination of the χ_i, say $f^\natural = \sum_{i=1}^n \lambda_i \chi_i$ $(\lambda_i \in \mathbb{C})$. Hence

$$\|f^\natural\|_\infty = \max_{1 \leq i \leq n} |\lambda_i|. \tag{1}$$

Next, we have

$$\|f^{\natural}\|_{\infty} \leq \|f\|_{\infty} \tag{2}$$

because for any $g \in G$,

$$|f^{\natural}(g)| \leq \int_K \int_K |f(g_1 k g_2)| \mathrm{d}k_1 \mathrm{d}k_2 \leq \|f\|_{\infty}.$$

Hence

$$\begin{aligned}|\mu(f)| = |\nu(f^{\natural})| &= \left|\sum_{i=1}^{n} \lambda_i \nu(\chi_i)\right| \\ &\leq \sum_{i=1}^{n} |\lambda_i||\nu(\chi_i)| \\ &\leq m_C \|f^{\natural}\|_{\infty} \leq m_C \|f\|_{\infty}\end{aligned}$$

by (1) and (2), where $m_C = \sum |\nu(\chi_i)|$ depends only on C. Hence μ is a measure on G, hence is a spherical measure. □

Proposition 1.2.3 *Let μ be a spherical measure on G. Then there exists a unique $\omega \in C(G, K)$ such that $\mathrm{d}\mu(g) = \omega(g^{-1})\mathrm{d}g$, i.e. such that*

$$\mu(f) = (f * \omega)(1) \quad \textit{for all } f \in C_c(G).$$

Proof Since $\mu \neq 0$ there exists $f_0 \in C(G, K)$ such that $\mu(f_0) \neq 0$. Let $f \in \mathcal{H}(G, K)$. Then

$$\mu(f * f_0) = \int_G f(g_1) \left(\int_G f_0(g_1^{-1} g) \mathrm{d}\mu(g) \right) \mathrm{d}g_1.$$

Since also $\mu(f * f_0) = \mu(f)\mu(f_0)$ it follows that

$$\mu(f) = \int_G f(g)\omega(g^{-1})\mathrm{d}g_1,$$

where

$$\omega(g_1) = \mu(f_0)^{-1} \int_G f_0(g_1^{-1} g)\mathrm{d}\mu(g). \tag{1.2.4}$$

Clearly the function ω so defined is continuous and bi-invariant with respect to K, i.e. $\omega \in C(G, K)$. If now $\phi \in C_c(G)$ then by Lemma 1.2.1

$$\begin{aligned}\mu(\phi) = \mu(\phi^{\natural}) &= \int_G \phi^{\natural}(g)\omega(g^{-1})\mathrm{d}g \\ &= \int_G \phi(g)\omega(g^{-1})\mathrm{d}g\end{aligned}$$

by the bi-invariance of ω. Hence $\mathrm{d}\mu(g) = \omega(g^{-1})\mathrm{d}g$, and since ω is continuous it is the only function with this property. □

Definition A function $\omega \in C(G, K)$ is a *zonal spherical function* (z.s.f for short) on G relative to K if $\omega(g^{-1})\mathrm{d}g$ is a spherical measure. The following proposition gives other equivalent definitions.

Proposition 1.2.5 *The following are equivalent, for a complex-valued function ω on G:*

(1) *ω is a z.s.f. on G relative to K;*
(2) *$\omega \in C(G, K)$; $\omega(1) = 1$; and $f * \omega = \lambda_f \omega$ for all $f \in \mathcal{H}(G, K)$, where λ_f is a scalar (depending on f);*
(3) *ω is continuous, not identically zero, and*

$$\omega(g_1)\omega(g_2) = \int_K \omega(g_1 k g_2)\mathrm{d}k \quad \textit{for all } g_1, g_2 \in G.$$

Proof

(1) $\Rightarrow$ (2). From the formula (1.2.4) it is clear that $\omega \in C(G, K)$ and $\omega(1) = 1$. Also, for all $f, f_0 \in \mathcal{H}(G, K)$ we have

$$\begin{aligned}\int_G f_0(g)(f * \omega)(g^{-1})\mathrm{d}g &= (f_0 * f * \omega)(1)\\ &= \mu(f_0 * f) = \mu(f)\mu(f_0),\end{aligned}$$

where μ is the associated spherical measure. Hence, putting $h = f * \omega - \mu(f)\omega$, we have

$$\int_G f_0(g)h(g^{-1})\mathrm{d}g = 0$$

for all $f_0 \in \mathcal{H}(G, K)$, and therefore $h = 0$ by Lemma 1.1.2.

(2) $\Rightarrow$ (3). We remark first that if $f \in \mathcal{H}(G, K)$ then

$$\int_G f(g)\omega(g^{-1})\mathrm{d}g = (f * \omega)(1) = \lambda_f \omega(1) = \lambda_f.$$

Next, we have

$$\begin{aligned}\lambda_f \omega(g') = (f * \omega)(g') &= \int_G f(g)\omega(g^{-1}g')\mathrm{d}g\\ &= \int_G f(g)\left(\int_K \omega(g^{-1}kg')\mathrm{d}k\right)\mathrm{d}g.\end{aligned}$$

Hence, if we put $\omega_{g'}(g) = \int_K \omega(gkg')\mathrm{d}k$, then $\omega_{g'} \in C(G, K)$ and

$$\int_G f(g)\omega_{g'}(g^{-1})\mathrm{d}g = \lambda_f \omega(g') = \int_G f(g)\omega(g^{-1})\omega(g')\mathrm{d}g.$$

Since this holds for all $f \in \mathcal{H}(G, K)$, it follows from Lemma 1.1.2 that

$$\omega(g^{-1})\omega(g') = \omega_{g'}(g^{-1}) = \int_K \omega(g^{-1}kg')\mathrm{d}k, \quad \forall g, g' \in G.$$

(3) $\Rightarrow$ (1). Let $\mu(f) = \int_G f(g)\omega(g^{-1})\mathrm{d}g$. We have to show that μ is a spherical measure. It is immediate from (3) that ω, and therefore μ, is bi-variant with respect to K. If $f_1, f_2 \in \mathcal{H}(G, K)$ then

$$\begin{aligned}
\mu(f_1 * f_2) &= \int_G \left(\int_G f_1(g_1)f_2(g_1^{-1}g)\mathrm{d}g_1\right)\omega(g^{-1})\mathrm{d}g \\
&= \int_G \int_G f_1(g_1)f_2(g_2)\omega(g_2^{-1}g_1^{-1})\mathrm{d}g_1\mathrm{d}g_2 \\
&= \int_G \int_G f_1(g_1)f_2(g_2)\left(\int_K \omega(g_2^{-1}kg_1^{-1})\mathrm{d}k\right)\mathrm{d}g_1\mathrm{d}g_2 \\
&= \int_G \int_G f_1(g_1)f_2(g_2)\omega(g_1^{-1})\omega(g_2^{-1})\mathrm{d}g_1\mathrm{d}g_2 \\
&= \mu(f_1) \cdot \mu(f_2).
\end{aligned}$$

□

We shall now forget about spherical measures and work only with the spherical functions, for which we may take (2) or (3) of Proposition 1.2.5 as equivalent definitions. In terms of spherical functions, Proposition 1.2.2 takes the form

Proposition 1.2.6 *If K is open in G, there is a one-one correspondence between the set of z.s.f. on G relative to K and the set of non-zero $\mathbb{C}$-algebra homomorphisms $\mathcal{H}(G, K) \to \mathbb{C}$: namely, the zonal spherical function ω corresponds to the homomorphism $\hat{\omega}\colon \mathcal{H}(G, K) \to \mathbb{C}$ defined by*

$$\hat{\omega}(f) = \int_G f(g)\omega(g^{-1})\mathrm{d}g.$$

Remark 1.2.7 If ω is a z.s.f. then so is the function $g \mapsto \omega(g^{-1})$. This is clear from (3) of Proposition 1.2.5 (since K, being compact, is unimodular). Also the constant function **1** on G is a z.s.f. More generally, any continuous homomorphism $\omega\colon G \to \mathbb{C}^*$ whose kernel contains K is a z.s.f. relative to K, again by Proposition 1.2.5 (3).

1.3 Induction of spherical functions

Proposition 1.3.1 *Let G_0 be a closed subgroup of G such that $G = KG_0$, and let ω_0 be a zonal spherical function on G_0 relative to $K_0 = K \cap G_0$. Define a function ϕ_0 on G by the rule $\phi(kg_0) = \omega_0(g_0)$, $k \in K, g_0 \in G_0$. Then the function defined by*

$$\omega(g) = \int_K \phi_0(gk)\mathrm{d}k$$

is a z.s.f. on G relative to K.

Proof We remark first that ϕ_0 is well defined, for if $kg_0 = k'g_0'$, $k' \in K, g_0' \in G_0$, then $k'^{-1}k = g_0'g^{-1} \in K_0$, so that $g_0' \in K_0 g_0$ and therefore $\omega_0(g_0') = \omega_0(g_0)$.

Clearly $\omega \in C(G,K)$ and $\omega(1) = 1$. Hence it is enough to show that $f * \omega$ is a scalar multiple of ω, for each $f \in \mathcal{H}(G,K)$.

From the definition of ϕ_0 we have $\phi_0(kg) = \phi_0(g)$ for all $k \in K$ and $g \in G$. Hence, if $g = kg_0' \in G$, $k \in K$ and $g_0, g_0' \in G_0$,

$$\begin{aligned}\int_{K_0} \phi_0(gk_0g_0)\mathrm{d}k_0 &= \int_{K_0} \phi_0(g_0'k_0g_0)\mathrm{d}k_0 \\ &= \int_{K_0} \omega_0(g_0'k_0g_0)\mathrm{d}k_0 \\ &= \omega_0(g_0')\omega_0(g_0) \text{ by Proposition 1.2.5}\end{aligned}$$

so that

$$\int_{K_0} \phi_0(gk_0g_0)\mathrm{d}k = \phi_0(g)\omega_0(g_0). \tag{3}$$

Hence, if $f \in \mathcal{H}(G,K)$ and $g = kg_0$,

$$\begin{aligned}(f * \phi_0)(g) &= \int_G f(kg_0g_1)\phi_0(g_1^{-1})\mathrm{d}g_1 \\ &= \int_G f(g_2)\phi_0(g_2^{-1}g_0)\mathrm{d}g_2 \quad (g_2 = g_0g_1) \\ &= \int_G f(g_2)\left(\int_{K_0} \phi_0(g_2^{-1}k_0g_0)\mathrm{d}k_0\right)\mathrm{d}g_2 \\ &= \int_G f(g_2)\phi_0(g_2^{-1})\omega_0(g_0)\mathrm{d}g_2 \text{ by (3) above} \\ &= \lambda_f\phi_0(g),\end{aligned} \tag{4}$$

where $\lambda_f = (f * \phi)(1)$. So, finally,

$$\begin{aligned}(f * \omega)(g) &= \int_G f(gg_1)\left(\int_K \phi_0(g_1^{-1}k)\mathrm{d}k\right)\mathrm{d}g_1 \\ &= \int_G f(g_2)\left(\int_K \phi_0(g_2^{-1}gk)\mathrm{d}k\right)\mathrm{d}g_2 \quad (g_2 = gg_1) \\ &= \int_K (f * \phi_0)(gk)\mathrm{d}k \\ &= \lambda_f \int_K \phi_0(gk)\mathrm{d}k \text{ by (4)} \\ &= \lambda_f\omega(g).\end{aligned}$$

□

The spherical function ω is said to be *induced* by ω_0.

Corollary 1.3.2 *With the hypotheses of Proposition* 1.3.1 *the function*

$$g \mapsto \int_K \phi_0(g^{-1}k)\mathrm{d}k$$

is a zonal spherical function on G.

Proof Proposition 1.3.1, Remark 1.2.7. □

Proposition 1.3.3 (Transitivity of induction) *Let $G_0 \subset G_1$ be closed subgroups of G such that $G = KG_0$. Let ω_0 be a z.s.f. on G_0 relative to $K_0 = K \cap G_0$, and let ω_1 be the z.s.f. on G_1 relative to $K_1 = K \cap G_1$ induced by ω_0. Then the z.s.f. on G relative to K induced by ω_0 and ω_1 are equal.*

Proof As in Proposition 1.3.1 define ϕ_0 by the rule

$$\phi_0(kg_0) = \omega_0(g_0), \quad k \in K, g_0 \in G_0,$$

and ϕ_1 by the rule

$$\phi_1(kg_1) = \omega_1(g_1) = \int_{K_1} \phi_0(g_1k_1)\mathrm{d}k_1, \quad k \in K, g_1 \in G_1.$$

Let ω be the z.s.f. on G relative to K induced by ω_1, so that

$$\omega(g) = \int_K \phi_1(gk)\mathrm{d}k.$$

Writing $gk = k'g_1(k)$ where $k' \in K$ and $g_1(k) \in G_1$, we have

$$\begin{aligned}
\omega(g) &= \int_K \omega_1(g_1(k))\mathrm{d}k \\
&= \int_K \left(\int_{K_1} \phi_0(g_1(k)k_1)\mathrm{d}k_1\right)\mathrm{d}k \\
&= \int_{K_1} \left(\int_K \phi_0\left(g_1(kk_1)\right)\right)\mathrm{d}k_1 \quad \text{since } g_1(kk_1) = g_1(k)k_1 \\
&= \int_K \phi_0(g_1(k))\mathrm{d}k \\
&= \int_K \phi_0(gk)\mathrm{d}k.
\end{aligned}$$

Hence ω is also the spherical function on G induced by ω_0. □

1.4 Positive definite spherical functions and representations of class one

Definition A function $\phi\colon G \to \mathbb{C}$ is said to be *positive definite* if ϕ is continuous, not identically zero, and if

$$\sum_{i,j=1}^{n} \lambda_i \overline{\lambda_j} \phi(g_i^{-1} g_j)$$

is real and ≥ 0 for all choices of $g_1, \ldots, g_n \in G$ and $\lambda_1, \ldots, \lambda_n \in \mathbb{C}$.

Lemma 1.4.1 *If ϕ is positive definite, then $\phi(1)$ is real and > 0; $|\phi(g)| \leq \phi(1)$ for all $g \in G$; and $\phi(g^{-1}) = \overline{\phi(g)}$.*

Proof Taking $n = 1$ in the definition we see that $\phi(1)$ is real and ≥ 0. Taking $n = 1$, the matrix

$$\begin{pmatrix} \phi(1) & \phi(g) \\ \phi(g^{-1}) & \phi(1) \end{pmatrix}$$

must be Hermitian and positive, whence $\phi(g^{-1}) = \overline{\phi(g)}$ and $|\phi(g)| \leq \phi(1)$. Since ϕ is not identically zero, it follows that $\phi(1) > 0$. □

Definitions. Let H be a separable complex Hilbert space, $B(\mathsf{H})$ the set of all continuous linear mappings $\mathsf{H} \to \mathsf{H}$. A *representation* T of G on H is a mapping $g \mapsto T_g$ of G into $B(\mathsf{H})$ such that

(1) $T_{g_1 g_2} = T_{g_1} \circ T_{g_2}$ for all $g_1, g_2 \in G$; $T_1 = 1$.
(2) For each $h \in \mathsf{H}$ the mapping $g \mapsto T_g(h)$ is a continuous map of G into H.

The representation T is *irreducible* if no closed subspace of H, other than $\{0\}$ and H, is invariant under T_g for all $g \in G$. T is *unitary* if each T_g is unitary.

Proposition 1.4.2

(i) *Let T be a unitary representation of G on H and let $h \in \mathsf{H}, h \neq 0$. Then the function ϕ defined by*

$$\phi(g) = \langle h, T_g(h) \rangle$$

(where $\langle\ ,\ \rangle$ is the scalar product on H) is positive definite.

(ii) *Conversely, let ϕ be a positive definite function on G. Then there exists a unitary representation T^{ϕ} of G on a Hilbert space H and a non-zero vector $h \in \mathsf{H}$ such that*

$$\phi(g) = \langle h, T_g^{\phi}(h) \rangle \quad \textit{for all } g \in G.$$

Proof

(i) is a straightforward verification: clearly ϕ is continuous and $\phi(1) \neq 0$; also

$$\begin{aligned}\sum_{i,j=1}^{n} \lambda_i \overline{\lambda_j} \phi(g_i^{-1} g_j) &= \sum_{i,j} \lambda_i \overline{\lambda_j} \langle h, T_{g_i}^{-1} T_{g_j}(h)\rangle \\ &= \left\langle \sum_i \lambda_i T_{g_i}(h), \sum_i \lambda_i T_{g_i}(h) \right\rangle \\ &= \left\| \sum \lambda_i T_{g_i}(h) \right\|^2\end{aligned}$$

which is real and ≥ 0.

(ii) We shall only sketch the proof. For more details see Helgason [10, p. 413]. Let V_ϕ be the subspace of $C(G)$ spanned by all the left-translates of ϕ, that is to say by all functions ${}^g\phi$ $(g \in G)$ defined by ${}^g\phi(g') = \phi(g^{-1}g')$. Define a scalar product on V_ϕ by

$$\langle {}^{g_1}\phi, {}^{g_2}\phi \rangle = \phi(g_1^{-1} g_2)$$

extended by linearity. The set $N = \{f \in V_\phi : \langle f, f \rangle = 0\}$ is a subspace of V_ϕ; the quotient V_ϕ / N inherits a scalar product from V_ϕ and hence can be completed to a Hilbert space H. Each $g \in G$ defines an endomorphism $f \mapsto {}^g f$ of V_ϕ which on passing to the quotient induces a unitary operator $T_g^\phi : \mathsf{H} \to \mathsf{H}$. One checks that T_ϕ is a representation of G. Finally, if $h \in \mathsf{H}$ is the image of $\phi \in V_\phi$, then

$$\langle h, T_g^\phi(h) \rangle = \langle \phi, {}^g\phi \rangle = \phi(g). \qquad \square$$

Corollary 1.4.3 *Suppose ϕ is a positive definite spherical function on G relative to K. Then there exists $h \in \mathsf{H}$ such that*

$$\phi(g) = \langle h, T_g^\phi(h) \rangle$$

for all $g \in G$ and such that $T_k^\phi(h) = h$ for all $k \in K$.

In fact, much more is true if we assume that G is unimodular and $\mathcal{H}(G, K)$ is a *commutative ring*. A representation T of G on a Hilbert space H is said to be of *class 1* (relative to the compact subgroup K) if T is irreducible and unitary and if there exists a non-zero vector $h \in \mathsf{H}$ such that $T_k(h) = h$ for all $k \in K$: that is to say, if the restriction of T to K contains the identity representation of K. Then there is the following result, which we shall not prove here:

Theorem 1.4.4 *Assume that G is unimodular and $\mathcal{H}(G, K)$ commutative.*

(i) *Let ω be a positive definite z.s.f. on G. Then the representation T^ω constructed from ω as in Proposition* 1.4.2 *is of class* 1.
(ii) *Conversely, if T is a representation of G of class 1 and if $h \in \mathsf{H}$ is unit vector left fixed by T_k for each $k \in K$, then h is unique up to a scalar multiple. Choosing h so that $\|H\| = 1$, the function*

$$\omega(g) = \langle h, T_g(h)\rangle$$

is a positive definite spherical function, and finally T is equivalent to T^ω.

We return to the situation of §1.3: G_0 is a closed subgroup of G and $G = KG_0$. Assume also that K is *open* in G, so that $K_0 = K \cap G_0$ is an open compact subgroup of G_0. Let $\mathrm{d}\xi$ be left Haar measure on G_0, normalized so that $\int_{K_0} \mathrm{d}\xi = 1$. The right Haar measure on G_0, normalized in the same way, will be of the form $\Delta(\xi)\mathrm{d}\xi$, where Δ is a continuous homomorphism of G_0 into the multiplicative group $\mathbb{R}_+^*$ of positive real numbers. Δ is trivial on K_0, because K_0 is compact.

If $\mathrm{d}k$ is normalised Haar measure on K then we have

$$\int_G f(g)\mathrm{d}g = \int_K \left(\int_{G_0} f(k\xi)\Delta(\xi)\right)\mathrm{d}k \tag{1.4.5}$$

for any integrable function f on G.

Lemma 1.4.6 *Let $\bar{K} = K/K_0$ and let g_1 be a fixed element of G. For $k \in K$, write*

$$g_1^{-1}k = k'\eta \quad k' \in K, \eta \in G_0.$$

Then the cosets $\bar{k}' = k'K_0$ and $K_0\eta K_0$ are uniquely determined by g_1 and $\bar{k} = kK_0$. If $\mathrm{d}\bar{k}$ is a relative invariant measure on $\bar{K}$, we have

$$\mathrm{d}\bar{k} = \Delta(\eta)\mathrm{d}\bar{k}'.$$

Proof Let $g = k\xi$ $k \in K, \xi \in G_0$. Then

$$g_1^{-1}g = g_1^{-1}k\xi = k'\eta\xi$$

so that $\mathrm{d}(g_1^{-1}g) = \mathrm{d}k'\Delta(\eta\xi)\mathrm{d}(\eta\xi) = \mathrm{d}k'\Delta(\eta)\Delta(\xi)\mathrm{d}\xi$ by Theorem 1.4.4; but also $\mathrm{d}(g_1^{-1}g) = \mathrm{d}g = \mathrm{d}k\Delta(\xi)\mathrm{d}\xi$. Hence the result. □

Proposition 1.4.7 *Let ω_0 be a z.s.f. on G_0 relative to K_0, and ω the induced z.s.f. on G relative to K (Proposition 1.3.1). Suppose that $\Delta^{1/2}\omega_0$ (i.e. $\xi \mapsto \Delta(\xi)^{1/2}\omega_0(\xi)$) is positive definite. Then ω is positive definite.*

Proof By Proposition 1.4.2 and Corollary 1.4.3 there exists a unitary representation T of G on a Hilbert space H_0 and a vector $h_0 \in \mathsf{H}_0$ such that

$$\Delta(\xi)^{1/2}\omega_0(\xi) = \langle h_0, T_\xi h_0\rangle$$

for all $\xi \in G_0$, and such that $T_{k_0}(h_0) = h_0$ for all $k_0 \in K_0$.

Let H be the Hilbert space of all classes of measurable functions $f\colon G \to \mathsf{H}_0$ which satisfy

(i) $f(g\xi) = \Delta(\xi)^{-1/2}T_\xi(f(g))$ for all $g \in G$ and $\xi \in G_0$;
(ii) $\int_K \|f(k)\|^2\mathrm{d}k < \infty$.

The inner product on H is

$$\langle f_1, f_2\rangle = \int_K \langle f_1(k), f_2(k)\rangle \mathrm{d}k.$$

Let $g_1 \in G$. For each $f \in \mathsf{H}$ define a function $U_{g_1} f$ on G by the rule

$$(U_{g_1} f)(g) = f(g_1^{-1} g).$$

Clearly $U_{g_1} f$ satisfies (i) above. Also we have

$$\|U_{g_1} f\|^2 = \int_K \|(U_{g_1} f)(k)\|^2 \mathrm{d}k = \int_K \|f(g_1^{-1}k)\|^2 \mathrm{d}k.$$

Now from (i) and the fact that Δ is trivial on K_0 it follows that (for fixed g_1) $\|f(g_1^{-1}k)\|^2$ depends only on the image of k in $\bar{K} = K/K_0$. Hence

$$\begin{aligned}
\|U_{g_1} f\|^2 &= \int_{\bar{K}} \|f(g_1^{-1}k)\}^2 \mathrm{d}\bar{k} \\
&= \int_{\bar{K}} \|f(k'\eta)\|^2 \Delta(\eta) \mathrm{d}\bar{k}' \quad \text{by Lemma 1.4.6} \\
&= \int_{\bar{K}} \|T_\eta f(k')\|^2 \mathrm{d}\bar{k}' \\
&= \|f\|^2
\end{aligned}$$

because T is unitary. Hence $U_{g_1} f \in \mathsf{H}$ and the operator U_{g_1} is unitary. (One can go on to show without any difficulty that U is a unitary representation of G, but we shall not need to use this fact.)

Let $h \in \mathsf{H}$ be the function $G \to \mathsf{H}_0$ defined by

$$h(k\xi) = \Delta(\xi)^{-1/2} T_\xi(h_0), \;\; k \in K, \xi \in G$$

(h is well defined because T/K_0 fixes h_0.) Now calculate $\langle h, U_g^{-1} h\rangle$:

$$\langle h, U_g^{-1} h\rangle = \int_K \langle h(k), h(gk)\rangle \mathrm{d}k \; :$$

putting $gk = k'\xi$ $(k' \in K, \xi \in G_0)$ this becomes

$$\int_K \langle h, \Delta(\xi)^{-1/2} T_\xi(h_0)\rangle \mathrm{d}k = \int_K \omega_0(\xi) \mathrm{d}k = \omega(g).$$

Hence $\omega(g) = \langle h, U_g^{-1} h\rangle$; since U_g is unitary, the calculation in Proposition 1.4.2 shows that ω is positive definite. □

Proposition 1.4.8 *Let ω be a z.s.f. on G relative to K, and suppose that ω is square-integrable. Then ω is positive definite.*

Proof G acts on $L^2(G)$ by left-translations: $(T_{g_0}f)(g) = f(g_0^{-1}g)$. Calculate the scalar product of ω and $T_{g_0}\omega$:

$$\begin{aligned}\langle \omega, T_{g_0}\omega\rangle &= \int_G \omega(g)\overline{\omega(g_0^{-1}g)}\mathrm{d}g \\ &= \int_G \omega(g_0 g)\overline{\omega(g)}\mathrm{d}g \\ &= \int_G \left(\int_K \omega(g_0 k g)\mathrm{d}k\right)\overline{\omega(g)}\mathrm{d}g \\ &= \int_G \omega(g_0)\omega(g)\overline{\omega(g)}\mathrm{d}g \quad \text{by Proposition 1.2.5} \\ &= \omega(g_0)\|\omega\|^2.\end{aligned}$$

Since T_{g_0} is unitary (by the left-invariance of the measure), it follows as in Proposition 1.4.2 that ω is positive definite. □

1.5 Fourier analysis

In this section we shall assume that G is unimodular and that the algebra $\mathcal{H}(G, K)$ is *commutative*. Let $L^1(G, K)$ (resp. $L^2(G, K)$) denote the space of integrable (resp. square-integrable) functions on G which are bi-invariant with respect to K.

Let Ω^+ denote the set of all positive definite z.s.f. on G relative to K. For each $f \in L^1(G, K)$, the *Fourier transform* of f is the function on Ω^+ defined by

$$\hat{f}(\omega) = \int_G f(g)\omega(g^{-1})\mathrm{d}g.$$

The set Ω^+ is given the weakest topology for which all the Fourier transforms of functions in $L^1(G, K)$ are continuous. Equivalently, the topology of Ω^+ is the compact-open topology: the sets $N(C, U) = \{\omega \in \Omega^+(C) \subset U\}$, where $C \subset G$ is compact and $U \subset \mathbb{C}$ is open, form a basis for the topology of Ω^+.

With this topology, Ω^+ is Hausdorff and locally compact, and the Fourier transforms of functions in $L^1(G, K)$ tend to 0 at infinity. We have

Ω^+ is compact $\Leftrightarrow$ K is open in G;
Ω^+ is discrete $\Leftrightarrow$ G is compact.

The fundamental result is

Theorem 1.5.1 (Plancherel–Godement theorem) *There exists a unique positive measure μ on Ω^+ such that*

(1) $f \in \mathcal{H}(G,K) \Rightarrow \hat{f} \in L^2(\Omega^+, \mu)$,
(2) $\int_G |f(g)|^2 \mathrm{d}g = \int_{\Omega^+} |\hat{f}(\omega)|^2 \mathrm{d}\mu(\omega)$ *for all* $f \in \mathcal{H}(G,K)$.

Moreover, the mapping $f \mapsto \hat{f}$ *extends to an isomorphism of Hilbert space* $L^2(G,K) \to L^2(\Omega^+, \mu)$.

This measure μ is called the *Plancherel measure* on Ω^+. The proof of Theorem 1.5.1 (or rather of a more general result which includes it as a special case) is sketched in [6] and given in detail in [4, Chapter XV].

From Theorem 1.5.1 there follows the usual duality of Fourier analysis, just as in the case of locally compact abelian groups. Convolution in $\mathcal{H}(G,K)$ becomes function multiplication on Ω^+: $\widehat{(f_1 * f_2)} = \hat{f}_1 \cdot \hat{f}_2$.

This theory includes as particular cases: (i) Fourier analysis on locally compact abelian groups; (ii) characters of compact groups. For (i), let G be abelian and $K = \{1\}$; then the functional equation Proposition 1.2.5 (3) shows that the z.s.f. on G in this case are just the continuous homomorphisms $G \to \mathbb{C}^*$, and since positive definite z.s.f. are bounded (by Lemma 1.4.1), they are just the characters of G. So in this case Ω^+ is the character group $\hat{G}$, and the Plancherel measure μ is the Haar measure on $\hat{G}$. For case (ii), let Γ be a compact group, take $G = \Gamma \times \Gamma$, and take K to be the diagonal subgroup. Then the bi-invariant functions on G may be identified with the class functions on Γ; all the z.s.f. on G relative to K are positive definite, and with the above identification they are precisely the functions $\gamma \mapsto \chi(\gamma)/\chi(1)$, where χ is an irreducible character of Γ.

For this and other examples (including $G = \mathrm{SO}_3(\mathbb{R})$, $K = \mathrm{SO}_2(\mathbb{R})$, which is the origin of the name "zonal spherical functions"), we refer to Helgason [10, Chapter X].

1.6 Notes on Chapter I

The material here is all standard. We have drawn largely on the account in [16].

1.7 Additions to Chapter I

Let $\mathbb{G}$ be a finite group, and let $\mathbb{H}$ be a subgroup of $\mathbb{G}$. We denote by $X := \mathbb{G}/\mathbb{H}$ the corresponding homogeneous space and by $C(X)$ the permutation module of all complex-valued functions defined on X. Then we have the isomorphism

$$\mathrm{Hom}_{\mathbb{G}}(C(X), C(X)) \simeq \bigoplus_{\pi \in \mathrm{Irr}(\mathbb{G})} \mathrm{M}_{d_\pi}(\mathbb{C}),$$

where d_π denotes the dimension of π and $\mathrm{M}_{d_\pi}(\mathbb{C})$ is the algebra of all $d_\pi \times d_\pi$ complex matrices.

The pair of groups $(\mathbb{G}, \mathbb{H})$ is called a *Gelfand pair* if the induced representation $\mathrm{Ind}_{\mathbb{H}}^{\mathbb{G}} 1$ is multiplicity free as a $\mathbb{G}$-module. In this situation there exists a unique $\mathbb{H}$-invariant element in each irreducible component of $\mathrm{Ind}_{\mathbb{H}}^{\mathbb{G}} 1$. These elements are called zonal spherical functions. There are interesting relations between zonal spherical functions on finite groups [Mac2] and hypergeometric functions [BI].

Mizukawa considered in [Miz] the Gelfand pair formed by the complex reflection group $G(r, 1, n)$ and the symmetric group S_n for n a positive integer, and expressed its zonal spherical functions in terms of multivariate hypergeometric functions.

Chapter II
Groups of p-adic type

2.1 Root systems

Let V be a real vector space of finite dimension $\ell > 0$, and $\langle\ ,\ \rangle$ a positive definite scalar product on V. Let V^* be the dual vector space. Then the scalar product defines an isomorphism of V onto V^*, and hence a scalar product $\langle\ ,\ \rangle$ on V^*. For each non-zero $v \in V^*$ let $v^\vee \in V$ be the image of $2v/\langle v, v\rangle$ under this isomorphism. Let h_a denote the kernel of a:

$$h_a = \{v \in V \ \ \langle a, v\rangle = 0\}\,,$$

and let s_a be the reflection in the hyperplane h_a:

$$s_a(v) = v - a(v)a^\vee, \quad v \in V.$$

We make s_a act also on V^*, by transposition: $s_a(b) = b \circ s_a$.

Definition A *root system* in V^* is a subset Σ_0 of V^* satisfying the following axioms:

(RS1) Σ_0 is finite, spans V^* and does not contain 0;
(RS2) $a \in \Sigma_0 \Rightarrow s_a(\Sigma_0) = \Sigma_0$;
(RS3) $a, b \in \Sigma_0 \Rightarrow a(b^\vee) \in \mathbb{Z}$.

We shall assume throughout that Σ_0 is *reduced*:

(RS4) If $a, b \in \Sigma_0$ are propositional, then $b = \pm a$;

and *irreducible*:

(RS5) If $\Sigma_0 = \Sigma_1 \cup \Sigma_2$ such that $\langle a_1, a_2\rangle = 0$ for all $a_1 \in \Sigma_1$ and $a_2 \in \Sigma_2$, then either Σ_1 or Σ_2 is empty.

The elements of Σ_0 are called *roots*. The group W_0 generated by the reflections s_a ($a \in \Sigma_0$) is a finite group of isometries of V, called the *Weyl group* of Σ_0. The connected components of the set $V - \bigcup_{a\in\Sigma_0} h_a$ are open simplicial cones called the

I. G. Macdonald and A.-M. Aubert, *Spherical Functions on a Group of p-adic Type*,
Lecture Notes in Mathematics 2392,
https://doi.org/10.1007/978-3-032-15671-6_2

chambers of Σ_0. These chambers are permuted simply transitively by W_0, and the closure of each chamber is a fundamental region for W_0.

Choose a chamber C_0. This choice determines a set Π_0 of l roots such that C_0 is the intersection of the open half-spaces $\{x \in V : a(x) > 0\}$ for $a \in \Pi_0$. The elements of Π_0 are called the *simple roots* (relative to the chamber C_0). A root b is said to be *positive* or *negative* according as it is positive or negative on C_0. Let Σ_0^+ (resp. Σ_0^-) denote the set of positive (resp. negative) roots.

Each root b can be expressed as a linear combination of the simple roots $b = \sum_{a \in \Pi_0} m_a a$ with coefficients $m_a \in \mathbb{Z}$ all of the same sign. The integer $\mathrm{ht}(b) := \sum_{a \in \Pi_0} m_a$ is called the *height* of b. There is a unique root (the "highest root") whose height is maximal.

Let $C_0^\perp$ denote the "obtuse cone"

$$C_0^\perp = \{v \in V : \langle v, v' \rangle \geq 0, \forall v' \in \overline{C_0}\}.$$

If $v \in \overline{C_0}$ and $w \in W_0$, then $v - wv \in C_0^\perp$.

A subset Φ_0 of Σ_0 is said to be *closed* if

$$a, b \in \Phi_0 \text{ and } a + b \in \Sigma_0 \Rightarrow a + b \in \Phi_0.$$

Suppose Φ_0 is closed and $\Phi_0 \cap -\Phi_0 = \emptyset$. Then there exists $w \in W_0$ such that $w\Phi_0 \subset \Sigma_0^+$.

The Weyl group W_0 is generated by the reflections s_a for $a \in \Pi_0$. Hence any $w \in W_0$ can be written as a product $w = s_1 \cdots s_r$, where $s_i = s_{a_i}$ and $a_i \in \Pi_0 (1 \leq i \leq r)$. If the number of factors s_i is as small as possible, then $s_1 \cdots s_r$ is a *reduced word* for w, and r is the *length* of w, written $\ell(w)$. The length of w is also equal to the number of positive roots b such that $w^{-1}b$ is negative. More precisely, these roots are $b_i = s_1 \cdots s_{i-1} a_i$ $(1 \leq i \leq r)$. (All this of course is relative to the choice of the chamber C_0.)

In particular, there exists a unique element $w_0 \in W_0$ which sends every positive root to a negative root. This leads to the following result, which will be useful later:

Lemma 2.1.1 *The positive roots can be arranged in a sequence* $(b_1, \ldots, b_n)$ *such that for each* $r = 0, 1, \ldots, n$ *the set of roots*

$$\{-b_1, -b_2, \ldots, -b_r, b_{r+1}, b_{r+2}, \ldots, b_n\}$$

is the set of positive roots relative to some chamber C_r.

Proof Let $s_1 s_2 \cdots s_n$ be a reduced word for w_0, where $s_i = s_{a_i}$, $a_i \in \Pi_0$, and as before put $b_i = s_1 s_2 \cdots s_{i-1} a_i$ $(1 \leq i \leq r)$. For each r, $s_{r+1} \cdots s_n$ is a reduced word, and therefore from the facts quoted above

$$\{a_{r+1}, s_{r+1} a_{r+2}, \ldots, s_{r+1} s_{r+2} \cdots s_{n-1} a_n\} = \Sigma_0^+ \cap (s_{r+1} \cdots s_n \Sigma_0^-).$$

Consequently

$$\{b_{r+1}, b_{r+2}, \ldots, b_n\} = (w_1 \cdots s_r \Sigma_0^+) \cap (s_0 \Sigma_0^-) = (s_1 \cdots s_r \Sigma_0^+) \cap \Sigma_0^+$$

and therefore

$$\{b_1, b_2, \ldots, b_r\} = (s_1 \cdots s_r \Sigma_0^-) \cap \Sigma_0^+$$

so that

$$\{-b_1, -b_2, \ldots, -b_r\} = (s_1 \cdots s_r \Sigma_0^+) \cap \Sigma_0^-$$

and finally

$$\{-b_1, -b_2, \ldots, -b_r, b_{r+1}, \ldots, b_n\} = s_1 \cdots s_r \Sigma_0^+$$

is the set of positive roots for the chamber $C_r = s_1 \cdots s_r C_0$. □

2.2 Affine roots

We retain the notation of §2.1. For each $a \in \Sigma_0$ and $m \in \mathbb{Z}$ we have an affine-linear function $a + m$ on V (namely, $x \mapsto a(x) + m$). Let

$$\Sigma := \{a + m \ : \ a \in \Sigma_0, n \in \mathbb{Z}\}.$$

The elements of Σ are called *affine roots*. We shall denote them by Greek letters $\alpha, \beta, \ldots$. If α is an affine root, so are $-\alpha$ and $\alpha + m$ for all integers m. If $\alpha = a + m$ then $a (\in \Sigma_0)$ is the *gradient* of α (in the usual sense of elementary calculus).

For each $\alpha \in \Sigma$, let h_α be the (affine) hyperplane on which α vanishes, and let s_α be the reflection in h_α. (From now on, only the affine structure and not the vector-space structure of V will come into play). The group W generated by the s_α as α runs through Σ is an infinite group of displacements of V, called the *Weyl group* of Σ or the *affine Weyl group*. W_0 is the subgroup of W which fixes the point 0. The translations belonging to W form a free abelian group Λ of rank ℓ, and W is the semi-direct product $W_0 \ltimes \Lambda$ of W_0 and Λ.

For each $a \in \Sigma_0$ let

$$t_a = s_a \circ s_{a+1} \in \Lambda.$$

The t_a $(a \in \Pi_0)$ are a basis of Λ. We have $t_a(0) = a^\vee$, and the mapping $\Lambda \to V$ defined by $t \mapsto t(0)$ maps Λ isomorphically onto the *coroot lattice*, which is the lattice spanned by the $a^\vee$ for $a \in \Sigma_0$.

The affine Weyl group acts on Σ by transposition: if $w \in W, \alpha \in \Sigma$ then $w(\alpha)$ is defined to be $\alpha \circ w^{-1}$.

The connected components of the set $V - \bigcup_{\alpha \in \Sigma} h_\alpha$ are open rectilinear ℓ-simplexes called the *chambers*[1] of Σ. The chambers are permuted *simply transitively* by W. The (non-empty) faces of the chambers are locally closed simplexes,

[1] Note that there is also the notion of *alcoves*.

called *facets*. Each point $v \in V$ lies in a unique facet F. The subgroup of W which fixes v also fixes each point of F.

Choose a chamber C. This choice determines a set Π of $\ell + 1$ affine roots, such that C is the intersection of the open half-spaces $\{v \in V : \alpha(v) > 0\}$ for $\alpha \in \Pi$. The affine roots belonging to Π are called the *simple affine roots* (relative to the chamber C).

The affine Weyl group W is generated by the relations s_α for $\alpha \in \Pi$. As in the case of W_0, we define the *length* $\ell(w)$ of an element $w \in W$. The length of w is also equal to the number of affine roots α which are positive on C and negative on wC, i.e. to the number of hyperplanes h_α which separate C and wC.

If F is a facet of C, then the subgroup of W which fixes a point v of F is generated by the s_α ($\alpha \in \Pi$) which fix v.

We shall need the following result later:

Lemma 2.2.1 *Let Φ be a set of affine roots whose gradients are all positive and distinct. Then there exists a translation $t \in \Lambda$ such that each $\alpha \in \phi$ is positive on $t(C_0)$.*

Proof Let the elements of Φ be $a + s_a$ ($a \in \Sigma_0^+$) and let $t \in \Lambda$. We have

$$(a + s_a)t(x) = a(x + t(0)) + s_a = a(x) + a(t(0)) + s_a.$$

If $v \in C_0$ then $a(v) > 0$ for all positive roots a. Hence $a + s_a$ will be positive on $t(C_0)$ provided that $a(t(0)) + s_a \geq 0$. Hence we have only to choose t so that $t(0)$ satisfies $a(t(0)) \geq -s_a$ for all $a \in \Sigma_0^+$ which occur as gradients of elements of Φ, and this is certainly possible. □

For later use, it will be convenient to choose C to be the unique chamber of Σ which is contained in the cone C_0 and has the origin 0 as one vertex. The following lemma presupposes this choice of C.

For each $w \in W$, let Φ_w be the set of affine roots which are positive on C and negative on wC.

Lemma 2.2.2 *Let w be an element of W such that $C \subset wC_0$. Then for all $w' \in W_0$ we have $\Phi_{ww'} = \Phi_w \cup w\Phi_{w'}$ (disjoint union).*

Proof If $\beta \in \Phi_w$ is positive on $ww'C$, then $\alpha = w^{-1}\beta$ is negative on C and positive on $w'C$, hence $\alpha(0) = 0$ (because the origin 0 belongs to the intersection of the closures of C and $w'C$). Consequently α is negative on C_0, hence β is negative on wC_0 and therefore on C: contradiction. Hence β must be negative on $ww'C_0$, i.e. $\beta \in \Phi_{ww'}$.

Next, let $\alpha \in \Phi_{w'}$ and put $\beta = w\alpha$. Then $\beta(wC_0) > 0$ and $\beta(ww'C_0) < 0$, hence β is positive on C and negative on $ww'C$, i.e. $\beta \in \Phi_{ww'}$. Hence $\Phi_w \cup w\Phi_{w'} \subset \Phi_{ww'}$.

Conversely, let $\beta \in \Phi_{ww'}$. If $\beta(wC) < 0$, then $\beta \in \Phi_w$; and if $\beta(wC) > 0$, then $w^{-1}\beta \in \Phi_w$. This establishes the opposite inclusion. Finally, the sets Φ_w and $w\Phi_{w'}$ are disjoint, because $\beta(wC) < 0$ for all $\beta \in \Phi_w$ and $\beta(wC) > 0$ for all $\beta \in w\Phi_{w'}$.□

Before going any further, it may be helpful to have a concrete example of the set-up so far:

Example 2.2.3 (Root system of type A_ℓ**)** Let V be the set of all $v = (v_0, \dots, v_\ell) \in \mathbb{R}^{\ell+1}$ such that $\sum_{i=1}^{\ell} v_i = 0$. The scalar product is $\langle v, v' \rangle = \sum v_i v'_i$. Let e_i be the i-th coordinate function on V ($e_i(v) = v_i$) and let

$$\Sigma_0 = \{e_i - e_j : i \neq j\}.$$

Then Σ_0 satisfies (RS1)–(RS5). The Weyl group W_0 acts by permuting the coordinates $v_0, \dots, v_\ell$ of $v \in V$, hence is isomorphic to the symmetric group $\mathfrak{S}_{\ell+1}$, of order $(\ell + 1)!$. We may take the chamber C_0 to be

$$C_0 = \{v \in V \;:\; v_0 > v_1 > \cdots > v_\ell\}$$

and the simple roots are then $e_{i-1} - e_i$ $(1 \leq i \leq l)$. The chamber[2] C is the open simplex

$$C = \{x \in V \;:\; 1 + v_\ell > v_0 > v_1 > \cdots > v_\ell\}.$$

2.3 BN-pairs

Definition[3] Let G be a group, B and N subgroups of G, and R a subset of the coset space $N/(B \cap N)$. The pair (B, N) is a *BN-pair* (or a *Tits system*[4]) in G if it satisfies the following four axioms:

(BN1) G is generated by B and N, and $H = B \cap N$ is normal in N.
(BN2) R generates the group $W = N/H$ and each $r \in R$ has order 2.
(BN3) $rBw \subset BwB \cup BrwB$ for all $r \subset R$ and $w \in W$.
(BN4) $rBr \not\subset B$, for each $r \in R$.

The group $W = N/H$ is the *Weyl group* of (B, N). The elements of W are cosets of H, and an expression such as BwB is to be read in the usual sense as a product of subsets of G. If S is a subset of W, then BSB means $\bigcup_{w \in S} BwB$.

One shows that the set R is uniquely determined by B and N: namely an element $w \in W$ belongs to R if and only if $B \cup BwB$ is a group. The elements of R are called the *distinguished generators* of W.

The axioms (BN1)–(BN4) have the following consequences. The proofs may be found in [1].

(2.3.1) *$G = BWB$, and the mapping $w \mapsto BwB$ is a bijection of W onto the double coset space $B \backslash G/B$.*

---

[2] This is an alcove.

[3] See also [1, §2, no. I].

[4] See [BT1, (1.2.6)].

(2.3.2) *For each subset S of R let W_S be the subgroup of W generated by S, and let $P_S = BW_SB$. Then P_S is a subgroup of G.*
(2.3.3) *The mapping $S \mapsto P_S$ is an inclusion-preserving bijection of the set of subsets of R onto the set of subgroups of G containing B.*
(2.3.4) *If S, S' are subsets of R, and P_S is conjugate to $P_{S'}$, then $S = S'$.*
(2.3.5) If S, S' are subsets of R, and $w \in W$, then $P_S w P_{S'} = BW_S w W_{S'} B$.
(2.3.6) *Each G_S is its own normalizer in G.*
(2.3.7) *If $r \in R$ and $w \in W$ and $\ell(rw) > \ell(w)$, then $BrwB = BrB \cdot BwB$.* (The *length* $\ell(w)$ of $w \in W$ is defined as the length of a reduced word in the generating set R, just as in §2.1 and §2.2.)

2.4 Buildings

Let G be a group with BN-pair (B, N) as in §2.3. We shall assume now that the Weyl group of (B, N) "is" an affine Weyl group W in the sense of §2.2. More precisely, we shall suppose we are given a reduced irreducible root system Σ_0 in a Euclidean space V, a chamber C for the associated affine root system Σ, and a homomorphism ν of N onto the affine Weyl group W such that

(1) $\mathrm{Ker}(\nu) = H$

so that we may identify (via ν) the Weyl group N/H of (B, N) with the Weyl group W of Σ;

(2) under this identification, the distinguished generators of N/H are the reflections in the walls of the chamber C; in other words, $R = \{s_\alpha : \alpha \in \Pi\}$.

The conjugates of B in G are called *Iwahori subgroups* of G. A *parahoric* subgroup of G is a *proper* subgroup containing an Iwahori subgroup. From the properties of BN-pairs recalled in §2.3, each parahoric subgroup P is conjugate to G_S for some proper subset S of R, and S is uniquely determined by P.

We set up a one-one correspondence $S \leftrightarrow F$ between the subsets S of R and the facets F of the chamber C as follows: To a facet F corresponds the set of all $s_\alpha \in S$ which fix F. In this correspondence, $\emptyset \leftrightarrow C$; $R \leftrightarrow \emptyset$; and "$S \leftrightarrow$ a vertex of C" $\Leftrightarrow S$ is a maximal proper subset of R.

If $S \leftrightarrow F$ we write G_F for G_S.

Each parahoric subgroup P determines uniquely a facet $F(P)$ of C: namely $F(P) = F \Leftrightarrow P$ is conjugate to G_F.

The *building* associated with the given BN-pair structure on G is the set $\mathcal{B}$ of all pairs (P, x) where P is a parahoric subgroup of G, and x is a point in the facet $F(P)$.

With each parahoric subgroup P we associate the subset $\mathcal{F}(P)$ of $\mathcal{B}$, where

$$\mathcal{F}(P) = \{(P, x) : x \in F(P)\}.$$

$\mathcal{F}(P)$ is called a *facet* of $\mathcal{B}$, of *type* $F(P)$. In particular, if P is an Iwahori subgroup, $\mathcal{F}(P)$ is a *chamber* of $\mathcal{B}$. If P is parahoric, there are only finitely many parahoric subgroups Q containing P, and we define

$$\overline{\mathcal{F}(P)} = \bigcup_{Q \supseteq P} \mathcal{F}(Q).$$

In this way the building $\mathcal{B}$ may be regarded as a geometrical simplicial complex, the $\mathcal{F}(P)$ being the "open" simplexes and the $\overline{\mathcal{F}(P)}$ the "closed' simplexes.

The group G acts on $\mathcal{B}$ by inner automorphisms: $g(P, x) = (gPg^{-1}, x)$.

2.4.1 Apartments

Consider the following subset $\mathcal{A}_0$ of $\mathcal{B}$:

$$\mathcal{A}_0 = \bigcup_{w \in W} \overline{\mathcal{F}(wBw^{-1})}.$$

Equivalently, $\mathcal{A}_0$ is the union of the facets $n\mathcal{F}(P)$ where $n \in N$ and $P \supset B$.

Remark Since we have identified N/H with W by means of the isomorphism induced by ν, expressions such as wPw^{-1} are meaningful, where $w \in W$ and P is a subgroup of G containing B. Hence W acts on $\mathcal{A}_0$: $w(P, x) = (wPw^{-1}, x) = (nPn^{-1}, x)$ for any $n \in N$ such that $\nu(n) = w$.

Proposition 2.4.1 *There exists a unique bijection* $j \colon A \to \mathcal{A}_0$ *such that*

(1) *for each facet* F *of* C *and each* $x \in F$,

$$j(x) = (G_F, x);$$

(2) $j \circ w = w \circ j$ *for all* $w \in W$.

Proof Let $y \in A$. Then y is congruent under W to a unique point x of $\overline{C}$: say $y = wx$ where $w \in W$ and $x \in F$, and F is a facet of the chamber C. Define $j(y)$ to be the point $(wG_F w^{-1}, x) \in \mathcal{A}_0$. We have to check that j is well defined. If also $y = w'x$, then $w^{-1}w'$ fixes x and therefore, by a property of reflection groups recalled in §2.2, belongs to the subgroup of W generated by the $s_\alpha, \alpha \in \Pi$, which fix x. In other words, $w^{-1}w' \in W_S$, where $S \leftrightarrow F$. Since $W_S \subset G_S = G_F$ it follows that $wG_F w^{-1} = w'G_F w'^{-1}$. Hence j is well defined, and clearly satisfies (1) and (2). The uniqueness of j is obvious. □

Lemma 2.4.2 *If* $g\mathcal{A}_0 = \mathcal{A}_0$ *then* $j^{-1} \circ (g|\mathcal{A}_0) \circ j \in W$.

Proof Let I_0 be the chamber $j(C) = \mathcal{F}(B)$. We have $g\mathrm{I}_0 = n_0\mathrm{I}_0$ for some $n_0 \in N$. Hence $g_0 = n_0^{-1}g$ fixes $\mathcal{A}_0$ as a whole and also fixes the chamber I_0 and all its facets; also it maps each facet to another facet of the same type. Transferring our attention to A, it follows that $\gamma = j^{-1} \circ (g_0|\mathcal{A}_0) \circ j$ maps A bijectively onto A and fixes the chamber C and all its facets. Hence γ also fixes the adjacent chambers

$s_\alpha C$ ($\alpha \in \Pi$), and so on. It follows that γ is the identity mapping and hence that $j^{-1} \circ (g|\mathcal{A}_0) \circ j = \nu(n_0) \in W$. □

The subcomplexes $g\mathcal{A}_0$ of $\mathcal{B}$, as g runs through G, are called the *apartments* of the building $\mathcal{B}$. If $\mathcal{A} = g\mathcal{A}_0$ is an apartment, we may transport the Euclidean structure of A to $\mathcal{A}$ via the bijection $(g|\mathcal{A}_0) \circ j$ and $(g'|\mathcal{A}_0) \circ j$ differ by an element of W, and hence we have a well-defined structure of Euclidean space on each apartment $\mathcal{A}$, and in particular a distance $d_{\mathcal{A}}(x, y)$ defined for $x, y \in \mathcal{A}$.

Thus the building $\mathcal{B}$ may be thought of as obtained by sticking together many copies of the Euclidean space V.

Lemma 2.4.3 *Any two facets of $\mathcal{B}$ are contained in a single apartment.*

Proof Consider two facets $\mathcal{F}(P_1)$ and $\mathcal{F}(P_2)$, where P_1, P_2 are parahoric subgroups of G: say $P_i = g_i G_{F_i} g_i^{-1}$ ($i = 1, 2$), where F_1, F_2 are facets of the chamber C in A. By (2.3.1) we can write $g_1^{-1} g_2 = b_1 n b_2$ with $b_1, b_2 \in B$ and $n \in N$. If $g = g_1 b_1$, then

$$P_1 = gG_{F_1}g^{-1}, \quad P_2 = g(nG_{F_2}n^{-1})g^{-1}$$

and therefore $\mathcal{F}(P_1)$ and $\mathcal{F}(P_2)$ are both in $g\mathcal{A}_0$. □

Lemma 2.4.4 *G is transitive on the set of pairs* $(\mathfrak{l}, \mathcal{A})$, *where $\mathcal{A}$ is an apartment and $\mathfrak{l}$ is a chamber in $\mathcal{A}$.*

Proof We have $\mathfrak{l} = \mathcal{F}(g_0 B g_0^{-1}) = g_0 \mathfrak{l}_0$ for some $g_0 \in G$, where $\mathfrak{l}_0 = \mathcal{F}(B)$. Hence we may assume that $\mathcal{B} = \mathcal{B}_0$. If $\mathcal{A} = g\mathcal{A}_0$ contains $\mathfrak{l}_0$, then $g^{-1}\mathfrak{l}_0$ is a chamber in $\mathcal{A}_0$ and hence $g^{-1}\mathfrak{l}_0 = n\mathfrak{l}_0$ for some $n \in N$. So if $g_1 = gn$ we have $\mathcal{A} = g_1\mathcal{A}_0$ and $\mathfrak{l}_0 = g_1\mathfrak{l}_0$. □

Proposition 2.4.5 *Let $\mathcal{A}, \mathcal{A}'$ be two apartments and let $\mathfrak{l}$ be a chamber contained in $\mathcal{A} \cap \mathcal{A}'$. Then there exists a unique bijection $\rho : \mathcal{A}' \to \mathcal{A}$ such that*

(1) *There exists $g \in G$ such that $\rho x = gx$ for all $x \in \mathcal{A}'$;*
(2) *$\rho x = x$ for all $x \in \mathfrak{l}$.*

Moreover, $\rho x = x$ for all $x \in \mathcal{A} \cap \mathcal{A}'$, and $d_{\mathcal{A}'}(x, y) = d_{\mathcal{A}}(\rho x, \rho y)$ for all $x, y \in \mathcal{A}'$.

Proof The existence of ρ follows immediately from Lemma 2.4.4. Uniqueness follows from Lemma 2.4.2: if $\rho_1, \rho_2 : \mathcal{A}' \to \mathcal{A}$ both satisfy (1) and (2), then $\rho_1 \circ \rho_2^{-1}$ fixes $\mathcal{A}$ as a whole and fixes the chamber $\mathfrak{l}$, hence is the identity map. That $d_{\mathcal{A}'}(x, y) = d_{\mathcal{A}}(\rho x, \rho y)$ follows from the definition of the metric on an apartment given after Lemma 2.4.2.

It remains to show that $\rho x = x$ for all $x \in \mathcal{A} \cap \mathcal{A}'$. Suppose $\mathcal{F} = \mathcal{F}(P)$ is a facet contained in $\mathcal{A} \cap \mathcal{A}'$. By Lemma 2.4.4 we may assume that $\mathcal{A}' = \mathcal{A}_0, \mathcal{A} = g\mathcal{A}_0$ and $\mathfrak{l} = \mathfrak{l}_0 = \mathcal{F}(B)$. Since $g\mathfrak{l}0 = \mathfrak{l}_0$ it follows that g normalizes B and therefore by (2.3.6) that $g \in B$.

Since $\mathcal{F}(P) \subseteq \mathcal{A}_0 \cap g\mathcal{A}_0$ we have

$$P = n_1 G_F n_1^{-1} = g(n_2 G_F n_2^{-1})g^{-1}$$

for some facet F of C and $n_1, n_2 \in N$. Since G_F is its own normalizer (2.3.6), we have $n_1^{-1} g n_2 \in G_F$ and therefore $Bn_2G_F = Bn_1G_F$. By (2.3.5), $Bn_2W_FB = Bn_1W_FB$ where W_F is the subgroup of W which fixes F. Hence (2.3.1) $n_2W_F = n_1W_F$; since $W_F \subset G_F$ it follows that $n_1G_Fn_1^{-1} = n_2G_Fn_2^{-1}$, hence $\mathcal{F} = g\mathcal{F} = \rho\mathcal{F}$. □

2.4.2 Retraction of the building onto an apartment

Theorem 2.4.6 *Let $\mathcal{A}$ be an apartment and $\mathfrak{l}$ a chamber in $\mathcal{A}$. Then there exists a unique mapping $\rho\colon \mathcal{B} \to \mathcal{A}$ such that, for all apartments $\mathcal{A}'$ containing $\mathfrak{l}$, $\rho|\mathcal{A}'$ is the bijection $\mathcal{A}' \to \mathcal{A}$ of Proposition* 2.4.5.

Proof Let $x \in \mathcal{B}$. By Lemma 2.4.3 there exists an apartment $\mathcal{A}_1$ containing x and $\mathfrak{l}$. Let $\rho_1 : \mathcal{A}_1 \to \mathcal{A}$ be the isomorphism of Proposition 2.4.5. If also x and $\mathfrak{l}$ are contained in another apartment $\mathcal{A}_2$, then we have also $\rho_2 : \mathcal{A}_2 \to \mathcal{A}$. But then $\rho_1^{-1} \circ \rho_2 : \mathcal{A}_2 \to \mathcal{A}_1$ is the isomorphism of (2.4.5) for the apartments $\mathcal{A}_1$ and $\mathcal{A}_2$ and the chamber $\mathfrak{l}$. Hence $\rho_1^{-1} \circ \rho_2$ fixes x, that is to say $\rho_1(x) = \rho_2(x)$. It follows that ρ as in the theorem is well defined, and the uniqueness is obvious. □

The mapping ρ of Theorem 2.4.6 is called the *retraction of $\mathcal{B}$ onto $\mathcal{A}$ with centre* $\mathfrak{l}$. It has the following properties:

Proposition 2.4.7

(1) $\rho x = x$ *for all* $x \in \mathcal{A}$.
(2) *For each facet $\mathcal{F}$ in $\mathcal{B}$, $\rho|\overline{\mathcal{F}}$ is an affine isometry of $\overline{\mathcal{F}}$ onto $\overline{\rho\mathcal{F}}$.*
(3) *If $x \in \overline{\mathfrak{l}}$, then $\rho^{-1}(x) = \{x\}$.*

Proof (1) and (2) are clear. As to (3), let $\mathcal{F}'$ be a facet such that $\mathcal{F} := \rho\mathcal{F}'$ is a facet of $\mathfrak{l}$. By Lemma 2.4.3 there exists an apartment $\mathcal{A}'$ containing $\mathfrak{l}$ and $\mathcal{F}'$, and $\rho|_{\mathcal{A}'}$ is a bijection of $\mathcal{A}'$ onto $\mathcal{A}$ leaving $\mathfrak{l}$ fixed. It follows that $\mathcal{F}' = \mathcal{F}$. □

Proposition 2.4.8

(i) *There exists a unique function $d\colon \mathcal{B} \times \mathcal{B} \to \mathbb{R}_+$ such that $d|_{\mathcal{A}\times\mathcal{A}}$ is the metric $d_{\mathcal{A}}$ for each apartment $\mathcal{A}$ in $\mathcal{B}$.*
(ii) *If ρ is a retraction of $\mathcal{B}$ onto an apartment $\mathcal{A}$ as in Theorem* 2.4.6, *then $d(\rho(x), \rho(y)) \leq d(x, y)$ for all $x, y \in \mathcal{B}$.*
(iii) *d is a G-invariant metric on $\mathcal{B}$.*

Proof

(i) Let $x, y \in \mathcal{B}_0$. By Lemma 2.4.3 there is an apartment $\mathcal{A}$ containing x and y. We wish to define $d(x, y)$ to be the Euclidean distance $d_{\mathcal{A}}(x, y)$ between x and y in $\mathcal{A}$. We must therefore show that, if x, y belong also to another apartment $\mathcal{A}'$, then $d_{\mathcal{A}}(x, y) = d_{\mathcal{A}'}(x, y)$.

Let $\mathfrak{l}$ be a chamber in $\mathcal{A}$ such that $x \in \overline{\mathfrak{l}}$ and let $\mathfrak{l}'$ be a chamber in $\mathcal{A}'$ such that $y \in \overline{\mathfrak{l}'}$. By Lemma 2.4.3 there is an apartment $\mathcal{A}''$ containing $\mathfrak{l}$ and $\mathfrak{l}'$. From Proposition 2.4.5 we have

$$d_{\mathcal{A}}(x, y) = d_{\mathcal{A}''}(x, y)$$

because $\mathcal{A}$ and $\mathcal{A}''$ have a chamber in common; and for the same reason

$$d_{\mathcal{A}'}(x, y) = d_{\mathcal{A}''}(x, y).$$

Hence d is well defined and is G-invariant (from the definition of $d_{\mathcal{A}}$).

(ii) Let $\mathcal{A}'$ be an apartment containing x and y. The image under ρ of the straight line segment joining x to y in $\mathcal{A}'$ will be a polygonal line in $\mathcal{A}$ of the same total length, by Proposition 2.4.7(ii), and endpoints $\rho(x)$ and $\rho(y)$. Hence the result.

(iii) Let $x, y, z \in \mathcal{B}$ and let $\mathcal{A}$ be an apartment containing x and y. Let ρ be a retraction of $\mathcal{B}$ onto $\mathcal{A}$. Then

$$d(x, y) \leq d(x, \rho(z)) + d(\rho(z), y) \leq d(x, z) + d(z, y)$$

by (ii) above, since $\rho(x) = x$ and $\rho(y) = y$. Hence d satisfies the triangle equality, and we have already observed that d is G-invariant. □

Proposition 2.4.9 *Let $x, y \in \mathcal{B}$. Then there is a unique geodesic joining x to y.*

Proof Let $\mathcal{A}$ be an apartment containing x and y and let $[xy]$ denote the straight line segment joining x to y in $\mathcal{A}$. Suppose z lies on a geodesic from x to y. Then $d(x, z) + d(z, y) = d(x, y)$. Let $z_0 \in [xy]$ be such that $d(x, z_0) = d(x, z)$ and hence also $d(z_0, y) = d(z, y)$. Let ρ be the retraction of $\mathcal{B}$ onto $\mathcal{A}$ with centre a chamber I such that $z_0 \in \bar{\mathrm{I}}$. Then we have

$$\begin{aligned} d(x, y) &\leq d(x, \rho(z)) + d(\rho(z), y) \\ &\leq d(x, z) + d(z, y) \quad \text{by Proposition 2.4.8(ii)} \\ &= d(x, y). \end{aligned}$$

Consequently

$$d(x, y) = d(x, \rho(z)) + d(\rho(z), y) \text{ and } d(x, \rho(z)) = d(x, z).$$

Since $x, y, \rho(z)$ are in the same Euclidean space $\mathcal{A}$ it follows that $\rho(z) = z_0$. Hence $z \in \rho^{-1}(z_0)$ and therefore $z = z_0$ by Proposition 2.4.7(iii). □

It follows from Proposition 2.4.9 that the straight line segment $[xy]$ lies in all apartments containing x and y.

Proposition 2.4.10 *$\mathcal{B}$ is complete with respect to the metric d.*

Proof Let (x_n) be a Cauchy sequence in $\mathcal{B}$. If ρ is a retraction of $\mathcal{B}$ onto an apartment $\mathcal{A}$, then $(\rho(x_n))$ is a Cauchy sequence in $\mathcal{A}$, by Proposition 2.4.8(ii), and therefore converges to a point $x \in \mathcal{A}$. The set of real numbers $d(x, g(x))$, where $g \in G$ is such that $x \neq gx$, has a positive lower bound μ, and we shall have

$$d(\rho(x_n), x) < \frac{\mu}{4}, \quad d(x_n, x_{n+1}) < \frac{\mu}{4}$$

for all sufficiently large n.

Now each $\rho(x_n)$ is of the form $g_n x_n$ for some $g_n \in G$. Put $y_n = g_n^{-1}x$. Then

$$\begin{aligned} d(y_n, y_{n+1}) &\leq d(y_n, x_n) + d(x_n, x_{n+1}) + d(x_{n+1}, y_{n+1}) \\ &= d(x, \rho(x_n)) + d(x_n, x_{n+1}) + d(\rho(x_{n+1}), x) \\ &< \frac{3\mu}{4} < \mu \end{aligned}$$

for all large n. Hence $y_n = y_{n+1} = \cdots = y$ and we have

$$d(x_n, y) = d(x_n, y_n) = d(\rho(x_n), x) \to 0$$

as $n \to \infty$. Hence $x_n \to y$. □

Fixed point theorem

Definition A subset X of $\mathcal{B}$ is *convex* if $x, y \in X \Rightarrow [xy] \subset X$.

Theorem 2.4.11 *Let X be a bounded non-empty subset of $\mathcal{B}$. Then the group of isometries γ of $\mathcal{B}$ such that $\gamma(X) \subset X$ has a fixed point in the closure of the convex hull of X.*

For the proof we shall require the following lemma:

Lemma 2.4.12 *Let $x, y, z \in \mathcal{B}$ and let m be the midpoint of $[xy]$. Then*

$$d(z, x)^2 + d(z, y)^2 \geq 2d(z, m)^2 + \frac{1}{2}d(x, y)^2.$$

Proof If x, y, z are all in the same apartment $\mathcal{A}$, this is true with equality, by elementary geometry (take z as origin). In the general case, let $\mathcal{A}$ be an apartment containing x and y and choose a chamber I in $\mathcal{A}$ such that $m \in \bar{\text{I}}$. Retract $\mathcal{B}$ onto $\mathcal{A}$ with centre I. Then

$$\begin{aligned} d(z, x)^2 + d(z, y)^2 &\geq d(\rho(z), x)^2 + d(\rho(z), y)^2 \\ &= 2d(\rho(z), m)^2 + \frac{1}{2}d(x, y)^2 \\ &= 2d(z, m)^2 + \frac{1}{2}d(x, y)^2. \end{aligned}$$

□

Proof (of Theorem* 2.4.11*) Let $\delta(X) = \sup_{x,y \in X} d(x, y)$ be the diameter of X. Choose a real number $k \in (0, 1)$ and let fX be the set of all $m \in \mathcal{B}$ such that m is the midpoint of $[xy]$ with $x, y \in X$ and $d(x, y) \geq k \cdot \delta(X)$. Clearly fX is non-empty and is contained in the convex hull of X.

If $m \in fX$ and $z \in X$, then by Lemma 2.4.12

$$\begin{aligned} d(z,m)^2 &\leq \frac{1}{2}d(z,x)^2 + \frac{1}{2}d(z,y)^2 - \frac{1}{4}d(x,y)^2 \\ &\leq \frac{1}{2}\delta(X)^2 + \frac{1}{2}\delta(X)^2 - \frac{k^2}{4}\delta(X)^2 \\ &= (1 - \frac{k^2}{4})\delta(X)^2 \end{aligned}$$

so that

$$d(z,m) \leq k_1\delta(X) \tag{5}$$

for some $k_1 < 1$.

If now $m \in fX$ and $z \in fX$, then again m is the midpoint of a segment $[xy]$ with $x, y \in X$ and $d(x,y) \geq k\delta(X)$, so that

$$\begin{aligned} d(z,m)^2 &\leq \frac{1}{2}d(z,x)^2 + \frac{1}{2}d(z,y)^2 - \frac{1}{4}d(x,y)^2 \\ &= (1 - \frac{k^2}{4})\delta(X)^2 - \frac{k^2}{4}\delta(X)^2 \\ &= (1 - \frac{k^2}{2})\delta(X)^2 \end{aligned}$$

so that

$$\delta(fX) \leq k_2\delta(X) \tag{6}$$

for some $k_2 < 1$. From (6) it follows that

$$\delta(f^nX) \to 0 \quad \text{as } n \to \infty. \tag{7}$$

For each n, pick $x_n \in f^nX$. Then by (5) and (6)

$$d(x_n, x_{n+1}) \leq k_1\delta(f^nX) \leq k_1k_2^n\delta(X).$$

Hence (x_n) is a Cauchy sequence in $\mathcal{B}$, hence by Proposition 2.4.10 converges to some $x \in \mathcal{B}$. Clearly x lies in the closure of the convex hull of X.

Now let γ be an isometry of $\mathcal{B}$ such that $\gamma X \subset X$. Then $\gamma f^nX \subset f^nX$ for all n. Let $x'_n = \gamma(x_n)$. Then (x'_n) is a Cauchy sequence with $x'_n \in f^nX$ for all n and converges to γx. But by (7) $d(x_n, x'_n) \to 0$ as $n \to \infty$. Hence $\gamma x = x$. Hence x is a fixed-point of the group of isometries γ of $\mathcal{B}$ such that $\gamma X \subset X$. □

Definition A subset M of G is said to be *bounded* if MX is bounded for all bounded subsets X of $\mathcal{B}$.

Lemma 2.4.13 *M is bounded $\Leftrightarrow$ M intersects only finitely many double cosets BwB $(w \in W)$.*

Proof Let ρ be the retraction of $\mathcal{B}$ onto the apartment $\mathcal{A}_0$ with centre $\mathrm{I}_0 = \mathcal{F}(B)$. Then $X \subset \mathcal{B}$ is bounded $\Leftrightarrow \rho X$ is bounded $\Leftrightarrow \rho X$ is contained in a finite union of closed chambers $\bar{\mathrm{I}}$ of $\mathcal{A}_0$. Hence M is bounded $\Leftrightarrow M\mathrm{I}$ is bounded, for each chamber $\mathrm{I} \subset \mathcal{A}_0 \Leftrightarrow M\mathrm{I}_0$ is bounded.

If $m \in M$, define $w_m \in W$ by the relation $m \in Bw_mB$. Then $M\mathrm{I}_0$ is bounded $\Leftrightarrow \bigcup_{m\in M} w_m\mathrm{I}_0$ is bounded $\Leftrightarrow$ the set $\{w_m : m \in M\}$ is finite $\Leftrightarrow M$ intersects only finitely many BwB. □

Theorem 2.4.14 *A subgroup Γ of G is bounded $\Leftrightarrow$ Γ is contained in a parahoric subgroup.*

Proof

$\Rightarrow$ Let $x \in \mathcal{B}$. Then $X = \Gamma x$ is bounded. Applying the fixed point theorem (Theorem 2.4.11), we see that there exists $y \in \mathcal{B}$ such that $\Gamma y = y$. If y lies in the facet $\mathcal{F}(P)$, then Γ normalizes the parahoric subgroup P and hence $\Gamma \subseteq P$ by (2.3.6).

$\Leftarrow$ Replacing Γ by a conjugate, we may assume that $\Gamma \subset G_S$, where S is a proper subset of $R = \{s_\alpha : \alpha \in \Pi\}$. Now $P_S = BW_SB$, and W_S is *finite* because $S \neq R$. Hence Γ is bounded by Lemma 2.4.13. □

Theorem 2.4.15 *Every bounded subgroup of G is contained in a maximal bounded subgroup. The maximal bounded subgroups are precisely the maximal parahoric subgroups of G, and form $\ell + 1$ conjugacy classes, corresponding to the vertices of the chamber C.*

2.5 Groups with affine root structure

2.5.1 Universal Chevalley groups

We retain the notation of §2.1 and §2.2. Let

$$L := \left\{u \in V^* \ : \ u(a^\vee) \in \mathbb{Z} \text{ for all } a \in \Sigma_0\right\}.$$

Then L is a lattice[5] of rank $\ell(= \dim V)$ in V^*, containing Σ_0.

Let k be any field, and k^+, $k^\times$ the additive and multiplicative groups of k. Associated with the pair (Σ_0, k) there is a *universal Chevalley group*[6] $G = G(\Sigma_0, k)$. We shall not go into the details of the construction of G, for which we refer to Steinberg [15, Chapter 6]. All we shall say here is that G is generated by elements $u_a(\xi)$ $(a \in \Sigma_0, \xi \in k)$ and contains elements $h(\chi)$ $(\chi \in \mathrm{Hom}(L, k^\times))$ with various relations between the u's and the h's. In particular, each u_a is an injective homomorphism of k^+ into G, and h is an injective homomorphism of $\mathrm{Hom}(L, k^\times)$ into G.

---

[5] L is the *weight lattice*.

[6] i.e., the group of k-rational points of a simply connected algebraic group.

Example 2.5.1 If Σ_0 is the root system of type A_ℓ (Example 2.2.3) then the universal Chevalley group $G(\Sigma_0, k)$ is the special linear group $SL_{\ell+1}(k)$. In this case the $u_a(\xi)$ and $h(\chi)$ may be taken as follows:

If $a = e_i - e_j$ $(i \neq j)$ then $u_a(\xi) = I + \xi E_{ij}$, where I is the unit matrix and E_{ij} is the matrix with 1 in the (i, j) place and 0 elsewhere. $h(\chi)$ is the diagonal matrix with $\chi(e_0), \ldots, \chi(e_\ell)$ in order down the diagonal.

Now suppose that the field k carries a discrete valuation val_k, i.e. a surjective homomorphism $\mathrm{val}_k : k^\times \to \mathbb{Z}$ such that $\mathrm{val}_k(\xi + \eta) \geq \min(\mathrm{val}_k(\xi), \mathrm{val}_k(\eta))$ for all $\xi, \eta \in k$ (conventionally $\mathrm{val}_k(0) = +\infty$). We consider certain subgroups of the Chevalley group $G = G(\Sigma_0, k)$.

First, let Z be the subgroup consisting of the $h(\chi)$ and let N be the normalizer of Z. Then there is a canonical isomorphism $N/Z \to W_0$, and one can show that this lifts uniquely to a homomorphism ν of N onto the affine Weyl group W. The restriction of ν to Z is described as follows. If $\chi \in \mathrm{Hom}(L, k^\times)$ then $\nu \circ \chi$ is a homomorphism $L \to Z$, which defines a linear form $V^* \to \mathbb{R}$, hence a vector in V. Then $\nu(h(\chi))$ is translation by this vector, and this translation belongs to W. We have $\nu(Z) = \Lambda$ and $Z = \nu^{-1}(\Lambda)$, where Λ is as before the translation subgroup of W. The kernel H of ν is the set of all $h(\chi)$ such that $\chi(L)$ is contained in the group of units of k.[7]

Next, for each $\alpha \in \Sigma$ we have a subgroup U_α of G, defined as follows. If $\alpha = a + m$, then

$$U_\alpha = \{u_a(\xi) : \xi \in k, \text{ and } \mathrm{val}_k(\xi) \geq m\}.$$

If Φ is any subset of Σ, we denote by U_Φ the subgroup of G generated by the U_α such that $\alpha \in \Phi$, and by G_Φ the subgroup generated by U_Φ and H. Then the triple $(N, \nu, (U_\alpha)_{\alpha\in\Sigma})$ has the following properties, for all $n \in N$ and $\alpha, \beta \in \Sigma$:

(I) $nU_\alpha n^{-1} = U_{\nu(n)\alpha}$.

From this it follows that H normalizes each U_α, hence each U_Φ. Consequently

$$G_\Phi = H \cdot U_\Phi = U_\Phi \cdot H.$$

(II) *If m is a positive integer, then $U_{\alpha+m}$ is a proper subgroup of U_α, and* $\bigcap_{m\geq 0} U_{\alpha+m} = \{1\}$.

From this it follows that for each $a \in \Sigma_0$, the union of the U_{a+r} $r \in \mathbb{Z}$ is a group, which we denote by $U_{(a)}$. (In the present situation $U_{(a)} = u_a(k^+)$.)

The next three properties relate to a pair of subgroups U_α and U_β in the following situations: (III) $\beta + \alpha$ is a positive constant; (IV) $\beta + \alpha = 0$; (V) neither $\beta + \alpha$ nor $\beta - \alpha$ is constant (i.e. the hyperplanes h_α, h_β are not parallel).

(III) *If $\beta = -\alpha + m$ with $m > 0$, then* $G_{\{\alpha,\beta\}} = U_\alpha H U_\beta$.

(IV) $G_{\{\alpha,-\alpha\}} = (U_\alpha \cdot \nu^{-1}(s_\alpha) \cdot U_\alpha) \cup (U_\alpha H U_{-\alpha+1})$.

---

[7] Here Z is the maximal split torus of $G = G(\Sigma_0, k)$, and $H = \mathrm{Ker}\,\nu = \{h(\chi) \in Z : \chi(L) \subseteq \mathfrak{o}_k^\times\}$ is the $\mathfrak{o}_k$-points of the torus Z. Thus we have an isomorphism $Z/H \simeq X_*(Z)$.

For $\alpha, \beta \in \Sigma$ let $[\alpha, \beta]$ denote the set of all γ of the form $m\alpha + n\beta$ with m, n positive integers, and let $[U_\alpha, U_\beta]$ be the subgroup of G generated by all commutators $(u, v) = uvu^{-1}v^{-1}$ with $u \in U_\alpha$ and $v \in U_\beta$. Then from Chevalley's commutator relations we have

(V) *For $\alpha, \beta \in \Sigma$, the commutator group $[U_\alpha, U_\beta]$ is contained in the subgroup generated by all the U_γ for $\gamma \in \Sigma$, and not parallel to α or β.*[8]

Let $U^+ = \langle U_{(a)} : a \in \Sigma_0^+ \rangle$ be the subgroup generated by the $U_{(a)}$ with a positive, and likewise $U^- = \langle U_{(a)} : a \in \Sigma_0^- \rangle$. Then

(VI) $U^+ \cap ZU^- = \{1\}$.

Finally,

(VII) G is generated by N and the U_α $(\alpha \in \Sigma)$.

2.5.2 Simply-connected simple algebraic groups

As before let k be a field with a discrete valuation val_k, and assume now that k is complete with respect to v, with *perfect* residue field.

Let $\mathbf{G}$ be a simply-connected simple linear algebraic group defined over k. Let $\mathbf{S}$ be a maximal k-split torus in $\mathbf{G}$, and let $\mathbf{Z}$, $\mathbf{N}$ be the centralizer and normalizer respectively of $\mathbf{S}$ in $\mathbf{G}$. Let $G = \mathbf{G}(k)$ be the group of k-rational points of $\mathbf{G}$. Then one of the main results of the structure theory of Bruhat and Tits ([BT1, BT2, BT3])[9] is:

Theorem 2.5.2 *There exists a reduced irreducible root system Σ_0, subgroups N and U_α $(\alpha \in \Sigma)$ of G, and a surjective homomorphism $\nu : N \to W$ such that the triple $(N, \nu, (U_\alpha)_{\alpha \in \Sigma})$ satisfies* (I)–(VII) *above.*

In fact one takes $N = \mathbf{N}(k)$, $Z = \mathbf{Z}(k)$, and $H = \mathrm{Ker}(\nu) = \mathbf{Z}(\mathfrak{o}_k)$ is the set of all $z \in Z$ such that $\mathrm{val}_k(\chi(z)) = 0$ for all characters χ of Z. W_0 is the relative Weyl group of $\mathbf{G}$ with respect to $\mathbf{S}$, but Σ_0 is not in general the relative root system, even if the latter is reduced. But every $a \in \Sigma_0$ is proportional to a root of G relative to $\mathbf{S}$, and conversely.

If $\mathbf{G}$ is split over k (i.e. if $\mathbf{S}$ is a maximal torus of $\mathbf{G}$) then we are back in the situation of (a) above.

If k is a non-Archimedean local field, (i.e., either k is a finite extension of the fields of p-adic numbers $\mathbb{Q}_p$ for some prime p or k is the field of formal Laurent series $\mathbb{F}_q((t))$) then $G = \mathbf{G}(k)$ inherits a topology from k for which G is a locally compact topological group. If K is a suitably chosen maximal compact subgroup of

---

[8] As pointed by P. Deligne to I. Macdonald, the original version of axiom (V) (i.e., "If h_α and h_β are not parallel, then $[U_\alpha, U_\beta] \subseteq U_{[\alpha,\beta]}$") was stronger than that provided by the theory of Bruhat and Tits expounded in [BT1, BT2], and not valid for the type C-B$_2$ in their classification.

[9] These replace the initial references [2, 3].

G, then as we shall see later (Corollary 3.3.7) the algebra $\mathcal{H}(G, K)$ is commutative, and the theory of zonal spherical functions on G relative to K is almost entirely a consequence of the properties (I)–(VII) listed above. We shall therefore take these properties as a set of axioms:

Definition 2.5.3 Let Σ_0 be a reduced irreducible root system and let Σ, W be the associated affine root system and affine Weyl group, as in §2.2. An affine root structure of type Σ_0 on a group G is a triple $(N, \nu, (U_\alpha)_{\alpha\in\Sigma})$, where N and the U_α are subgroups of G, and ν is a homomorphism of N onto W, satisfying (I)–(VII) above.

2.6 Cartan and Iwasawa decompositions

In this section G is a group endowed with an affine root structure $(N, \nu, (U_\alpha)_{\alpha\in\Sigma})$ of type Σ_0 (2.5.3). We shall develop some consequences of the axioms (I)–(VII).

Proposition 2.6.1 *Let Φ be a subset of Σ such that*

(1) *if $\alpha, \beta \in \Phi$ and $\alpha + \beta \in \Sigma$, then $\alpha + \beta - m \in \Phi$ for some $m \geq 0$;*
(2) *if $\alpha, \beta \in \Phi$ and $\alpha \neq \beta$, then the hyperplanes h_α, h_β are not parallel.*

Let U_Φ be the subgroup of G generated by the $U_\alpha, \alpha \in \Phi$. Then every element of U_Φ can be written uniquely in the form $u_1 \cdots u_r$, where $u_i \in U_{\alpha_i}$ $(1 \leq i \leq r)$ and $\alpha_1, \dots, \alpha_r$ are the elements of Φ in any fixed order.

Proof Let a_i be the gradient of α_i. Then $a_1, \dots, a_r$ are distinct by (2) and $\Phi_0 = \{a_1, \dots, a_r\}$ is a *closed* subset of Σ_0 such that $\Phi_0 \cap -\Phi_0 = \emptyset$ (by (1) and (2)). Hence there exists $w \in W_0$ such that $w\Phi_0 \subseteq \Sigma_0^+$ (2.2.1). Choosing $n \in N$ such that $\nu(n) = w$, we have $nU_\Phi n^{-1} = U_{w\Phi}$ by (I); hence we may assume without loss of generality that $\Phi_0 \subseteq \Sigma_0^+$.

Assume first that $(a_1, \dots, a_r)$ in this order is a subsequence of the sequence $(b_1, \dots, b_n)$ of (2.1.1). Let $\Phi_1 = \{\alpha_2, \dots, \alpha_r\}$. If m, n are positive integers and $i > 1$, then $ma_1 + na_i \neq a_1$, hence $U_{[\alpha_1,\alpha_i]} \subseteq U_{\Phi_1}$. Now use the commutator axiom (V): if $u_1 \in U_{\alpha_1}$ and $u_i \in U_{\alpha_i}$ $(i > 1)$ we have $u_1u_iu_1^{-1}u_i^{-1} \in U_{[\alpha_1,\alpha_i]}$ and therefore $u_1u_iu_1^{-1} \in U_{\Phi_1}$. This shows that U_{Φ_1} is normal in U_Φ, and hence that $U_\Phi = U_{\alpha_1}U_{\alpha_1}$. By induction on $r = \mathrm{Card}(\Phi)$ we conclude that

$$U_\Phi = U_{\alpha_1}U_{\alpha_2} \cdots U_{\alpha_r}.$$

Next we shall show that every $u \in U_\Phi$ is uniquely of the form $u = u_1 \cdots u_r$ with $u_i \in U_{\alpha_i}$ $(1 \leq i \leq r)$. Again by induction, it is enough to show that u_1 is uniquely determined by u, and for this it is enough to show that $U_{\alpha_1} \cap U_{\Phi_1} = \{1\}$. By (2.1.1) we can assume that a_1 is negative and $a_2, \dots, a_r$ are positive. Then $U_{\alpha_1} \subset U^-$ and $U_{\Phi_1} \subset U^+$, hence the result follows from (VI). This proves (2.6.1) for the particular ordering chosen. For an arbitrary ordering of Φ it follows from (2.6.2) below, whose proof we leave to the reader.

Lemma 2.6.2 *Let X be a group with subgroups $X_1, \ldots, X_r$ such that*

(1) $X = X_1 X_2 \cdots X_r$ with uniqueness of expression;
(2) $X_i X_{i+1} \cdots X_r$ is a normal subgroup of X, for $i = 1, 2, \ldots, r$.

If p is any permutation of $\{1, 2, \ldots, r\}$, then $X = X_{p(1)} X_{p(2)} \cdots X_{p(r)}$ with uniqueness of expression. □

Proposition 2.6.3 $U^+ = \prod_{a \in \Sigma_0^+} U_{(a)}$ *with uniqueness of expression, the positive roots being taken in any fixed order. Likewise* $U^- = \prod_{a \in \Sigma_0^-} U_{(a)}$.

Proof Arrange the positive roots in a sequence $(b_1, \ldots, b_n)$ satisfying (2.1.1). Follow the proof of (2.6.1) to show that $U^+ = \prod_{i=1}^n U_{b_i}$ with uniqueness of expression, and then use (2.6.2). □

Notation If S is any subset of V, let (S) denote the set of all $\alpha \in \Sigma$ which are ≥ 0 on S. Then the subgroups $U_{(S)}, G_{(S)}$ are defined.

Let $a \in \Sigma_0$. If there is an affine root with gradient a belonging to (S), there is a least such affine root, say $\alpha : \alpha \in (S)$ and $\alpha - 1 \notin (S)$. Let $U_{(S),a}$ denote the corresponding subgroup U_α of G. If there is no affine root with gradient a belonging to (S), define $U_{(S),a}$ to be $\{1\}$. Let $U_{(S)}^+$ (resp. $U_{(S)}^-$) be the group generated by the $U_{(S),a}$ with $a \in \Sigma_0^+$ (resp. $a \in \Sigma_0^-$).

Proposition 2.6.4 *Suppose S has non-empty interior. Then* $G_{(S)} = U_{(S)}^+ H U_{(S)}^-$.

Proof By (2.6.1), $U_{(S)}^\pm = \prod_{a \in \Sigma_0^\pm} U_{(S),a}$, the product being taken in any order. Since $G_{(S)}$ is generated by the $U_{(S),a}$ and H, it is enough to show that $U_{(S)}^+ H U_{(S)}^-$ is a group. For this it is enough to show that $U_{(S)}^+ H U_{(S)}^- = U_{(S)}^- H U_{(S)}^+$.

Arrange the positive roots in a sequence $(b_1, \ldots, b_n)$ satisfying (2.1.1). Let $U_{\pm r} = U_{(S), \pm b_r}$ $(1 \leq r \leq n)$, and let V_r^+ (resp. V_r^-) be the subgroup of G generated by $U_{-1}, \ldots, U_{-r}, U_{r+1}, \ldots, U_n$ (resp. $U_1, \ldots, U_r, U_{-r-1}, \ldots, U_{-n}$). Since S has non-empty interior it follows from (III) that $U_r H U_{-r}$ is a group and therefore $U_r H U_{-r} = U_{-r} H U_r$.

Hence, by (2.6.1), we have $V_{r-1}^+ H V_{r-1}^- = V_r^+ H V_r^-$ for $r = 1, 2, \ldots, n$. Consequently

$$U_{(S)}^+ H U_{(S)}^- = V_0^+ H V_0^- = V_n^+ H V_n^- = U_{(S)}^- H U_{(S)}^+.$$ □

Corollary 2.6.5 *If $\alpha \in \Sigma$ then $U_\alpha \subseteq G_{(S)} \Leftrightarrow \alpha \in (S)$.*

Proof Suppose $U_\alpha \subseteq G_{(S)}$. Let $u \in U_\alpha$, then $u = u^+ h u^-$ with $u^\pm = U_{(S)}^\pm$ and $h \in H$. Suppose, to fix the ideas, that the gradient a of α is a positive root. Then $(u^+)^{-1} u = h u^- \in U^+ \cap Z U^- = \{1\}$ by (VI). Hence $u = u^+$, i.e. $u \in U_{(S)}^+$. By (2.6.1) and (2.6.3) it follows that $u \in U_{(S),a}$. Hence $U_\alpha \subseteq U_{(S),a}$ and therefore by (II) we have $\alpha \in (S)$. The reverse implication is obvious. □

We shall now show that the group G carries two (B, N)-pair structures, one with Weyl group W (2.6.8) and the other with Weyl group W_0 (2.6.9). To establish these results we need a couple of preliminary lemmas.

Lemma 2.6.6

(1) *If* $\alpha \in \Sigma$, *then* $U_{-\alpha} - U_{-\alpha+1} \subset U_\alpha \cdot \nu^{-1}(s_\alpha) \cdot U_\alpha$.
(2) *If* $a \in \Sigma_0$, *then* $U_{-a} \subset U_a \cdot \nu_0^{-1}(s_a) \cdot U_a$.

(Here ν_0 is the epimorphism $N \longrightarrow W_0$ obtained by composing $\nu : N \longrightarrow W$ with the epimorphism $W \longrightarrow W_0$ defined by the semi-direct decomposition $W = \Lambda \rtimes W_0$. *Notice the difference between* U_α *and* U_a.)

Proof

(1) Let $u \in U_{-\alpha}$, then by (IV) we have

$$u \in U_\alpha \cdot \nu^{-1}(s_\alpha) \cdot U_\alpha \cup U_\alpha H U_{-\alpha+1}.$$

Suppose that $u \in U_\alpha H U_{-\alpha+1}$. Then $u = u_1 h u_2$ with $u \in U_\alpha, u_2 \in U_{-\alpha+1}, h \in H$. Hence $uu_2^{-1} = u_1 h \in U_\alpha H \cap U_{-\alpha} = \{1\}$ by (VI). Consequently $u = u_2 \in U_{-\alpha+1}$, whence the result.

(2) Let $u \in U_{-a}, u \neq 1$. Then by (II) there exists an affine root α with gradient a such that $u \in U_{-\alpha} - U_{-\alpha+1}$. Hence by (1) just proved we have

$$u \in U_\alpha \cdot \nu^{-1}(s_\alpha) \cdot U_\alpha \subset U_a \cdot \nu_0^{-1}(s_a) \cdot U_a. \qquad \square$$

Let $B = G_{(C)}$ and let $B_0^{\pm} = ZU^{\pm}$. Since Z normalizes U^+, B_0^+ is a subgroup of G, and is the semidirect product of Z and U^+ by axiom (VI). Similarly for B_0^-.

Lemma 2.6.7

(1) *If* $\alpha \in \Pi$ *and* $n_\alpha \in \nu^{-1}(s_\alpha)$, *then* $Bn_\alpha B = Bn_\alpha U_\alpha$.
(2) *If* $a \in \Pi_0$ *and* $n_a \in \nu^{-1}(s_a)$, *then* $B_0^+ n_a B_0^+ = B_0^+ n_a U_a$.

Proof

(1) Since $\alpha(C) > 0$ we have $U_\alpha \subset B$, hence $Bn_\alpha U_\alpha \subseteq Bn_\alpha B$. Conversely, if $\beta \in (C)$ and $\beta \neq \alpha$, then since h_α is the only hyperplane separating C and $s_\alpha C$, it follows that β is positive on $s_\alpha C$, hence that $U_\beta \subset G_{(s_\alpha C)} = n_\alpha^{-1} B n_\alpha$. Consequently $Bn_\alpha U_\beta = Bn_\alpha$ for all such β, and therefore $Bn_\alpha B = Bn_\alpha U_\alpha$.

(2) Let U^a be the group generated by the U_b with $b \in \Sigma_0^+$ and $b \neq a$. By (2.6.3) we have $U^+ = U^a U_a$. Hence, since s_a permutes $\Sigma_0^+ - \{a\}$,

$$B_0^+ n_a B_0^+ = B_0^+ n_a Z U^a U_a = B_0^+ Z U^a n_a U_a = B_0^+ n_a U_a. \qquad \square$$

Theorem 2.6.8 *Let* $B = G_{(C)}$. *Then* (B, N) *is a BN-pair in* G. *We have* $B \cap N = H$, *so that* ν *induces an isomorphism of* $N/(B \cap N)$ *onto to the affine Weyl group* W. *The distinguished generators of* $N/(B \cap N)$ *correspond under this isomorphism to the reflections in the walls of the chamber* C.

Proof

(1) Let $n \in B \cap N$. Then $nG_{(C)}n^{-1} = G_{(C)}$, so that $G_{(wC)} = G_{(C)}$ where $w = \nu(n)$. Suppose that $w \neq 1$, so that $wC \neq C$. Then there exists a hyperplane

h_α ($\alpha \in \Sigma$) separating C and wC, hence an affine root α which is positive on wC and negative on C. Then $U_\alpha \subset G_{(wC)} = G_{(C)}$, and since $\alpha \notin (C)$ this contradicts (2.6.5). Hence $w = 1$ and therefore $n \in H := \operatorname{Ker} \nu$. Hence $B \cap N = H$.

(2) We have to verify the axioms (BN1)–(BN4) of §2.3. Of course, (BN1) follows immediately from (I) and (VII). (BN2) is clear: we define R to be the set $\{\nu^{-1}(s_\alpha) : \alpha \in \Pi\}$. So is (BN4), because we have already shown that $G_{(wC)} \neq G_{(C)}$ if $w \neq 1$. As to (BN3), let $w \in W$ and $\alpha \in \Pi$, and let $n \in \nu^{-1}(w), n_\alpha \in \nu^{-1}(s_\alpha)$. Then

$$n_\alpha Bn \subset Bn_\alpha Bn = Bn_\alpha U_\alpha n$$

by (2.6.7). Suppose first that $\alpha(wC) > 0$. Then $U_{w^{-1}\alpha} \subset B$ and so

$$Bn_\alpha U_\alpha n = Bn_\alpha n U_{w^{-1}\alpha} \subset Bn_\alpha nB.$$

If on the other hand $\alpha(wC) < 0$, then

$$Bn_\alpha U_\alpha n = BU_{-\alpha} n_\alpha n$$

and by axiom (IV) we have

$$U_{-\alpha} \subset U_\alpha n_\alpha HU_\alpha \cup U_\alpha HU_{-\alpha+1},$$

so that

$$\begin{aligned} BU_{-\alpha} n_\alpha n &\subset Bn_\alpha U_\alpha n_\alpha n \cup BU_{-\alpha+1} n_\alpha n \\ &= BU_{-\alpha} n \cup Bn_\alpha n, \quad \text{since } -\alpha + 1 \in (C) \\ &= BnU_{-w^{-1}\alpha} \cup Bn_\alpha n \\ &\subset BnB \cup Bn_\alpha nB, \end{aligned}$$

since $-w^{-1}\alpha \subset (C)$. Hence $n_\alpha Bn \subset BnB \cup Bn_\alpha nB$ in all cases. □

Definition It follows that all the results of §2.4 are applicable in the present situation. The group $B = G_{(C)}$ and its conjugates are the *Iwahori subgroups* of G; the groups $G_{(F)}$ and their conjugates, where F is a facet of C, are the *parahoric subgroups*.

Theorem 2.6.9 *Let $B_0^+ = ZU^+$. Then (B_0^+, N) is a BN-pair in G. We have $B_0^+ \cap N = Z$, so that ν_0 induces an isomorphism of $N/(B_0^+ \cap N)$ onto the Weyl group W_0. The distinguished generators of $N/(B_0^+ \cap N)$ correspond under this isomorphism to the reflections in the walls of the cone C_0.*

Proof (1) Let $n \in B_0^+ \cap N$. Then $n = z_0 u^+$ ($z_0 \in Z, u \in U^+$) and therefore $n_0 = z_0^{-1} n \in N \cap U^+$. So we have $n_0 = \prod_{a \in \Sigma_0^+} u_a$ with $u_a \in U_a$ by (2.6.3). For each $a \in \Sigma_0^+$, pick an affine root α with gradient a such that $u_a \in U_\alpha$. Applying (2.2.1) to this set of affine roots, we see that $n_0 \in U_{(tC_0)}$ for some translation $t \in \Lambda$, and hence $n_1 = z^{-1} n_0 z \in U_{(C_0)}$ for some $z \in Z$. We have $n_1 \in N \cap U_{(C_0)} \subset N \cap B = H$ by (2.6.8). Hence $n_0 \in H \cap U^+ = \{1\}$ by axiom (VI). So $n = z_0 \in Z$ and therefore $B_0^+ \cap N = Z$.

(2) We have to verify the BN-axioms. As before, (BN1) follows immediately from (I) and (VII). (BN2): take the distinguished generators of N/Z to be $\nu_0^{-1}(s_a)$, $a \in \Pi_0$. As to (BN4), let $a \in \Pi_0$ and let $n_a \in \nu_0^{-1}(s_a)$, then if $B_0^+ = n_a B_0^+ n_a^{-1}$ we have

$$\begin{aligned} B_0 = B_0^+ n_a B_0^+ n_a^{-1} &= B_0^+ n_a U_a n_a^{-1} \quad \text{by (2.6.7)} \\ &= B_0^+ U_{-a} \end{aligned}$$

so that $U_{-a} \subset B_0 \cap U^- = ZU^+ \cap U^- = \{1\}$ by axiom (VI). Since $U_{-a} \neq \{1\}$ by axiom (II), we have a contradiction.

Finally, (BN3). Let $w \in W_0$, $a \in \Pi_0$, $n \in \nu_0^{-1}(w)$, $n_a \in \nu_0^{-1}(s_a)$. Then $n_a B_0^+ n \subset B_0^+ n_a B_0^+ n = B_0^+ n_a U_a n$ by (2.6.7).

Suppose first that $w^{-1}a$ is positive, then[10]

$$B_0^+ n_a U_a n = B_0^+ n_a n U_{w^{-1}a} \subset B_0^+ n_a n B_0^+.$$

If on the other hand $w^{-1}a$ is negative, then[11]

$$\begin{aligned} B_0^+ n_a U_a n &= B_0^+ U_{-a} n_a n \\ &\subset B_0^+ n_a U_a n_a n \quad \text{by (2.6.6)} \\ &= B_0^+ U_{-a} n = B_0^+ n U_{-w^{-1}a} \subset B_0^+ n B_0^+. \end{aligned}$$ □

Remark Of course there is a corresponding result for $B_0^- = ZU^-$.

Proposition 2.6.10 $G = BNB_0^+$.

Proof We remark first that G is generated by B_0^+ and by the U_{-a} ($a \in \Pi_0$). For let $G' = \langle B_0^+, (U_{-a})_{a \in \Pi_0} \rangle$, and let α be an affine root with gradient $a \in \Pi_0$. Then G' contains the group $\langle U_\alpha, U_{-\alpha}, H \rangle$ and hence by axiom (IV) contains $\nu^{-1}(s_\alpha)$. Since G' contains Z it follows that $N \subset G'$, hence that $U_\beta \subset G'$ for all $\beta \in \Sigma$. Hence $G' = G$.

Next, let $a \in \Pi_0$ and let C' be any chamber in A. Then

$$U_{-a} \subset G_{(C')} N U_a. \tag{8}$$

For let α be an affine root with gradient a. If $-\alpha \in (C')$ we have $U_{-\alpha} \subset G_{(C')}$. If on the other hand $\alpha \in (C')$, then

$$\begin{aligned} U_{-\alpha} - U_{-\alpha+1} &\subset U_\alpha \cdot \nu^{-1}(s_\alpha) \cdot U_\alpha \quad \text{by (2.6.6)} \\ &\subset G_{(C')} N U_a. \end{aligned}$$

[10] Here we use $nU_{w^{-1}a}n^{-1} = U_a$.

[11] Here we use $n_a U_a n_a^{-1} = U_{-a}$.

This proves (8), from which it follows that $U_{-\alpha} \subset nBNU_a$ for all $n \in N$, so that

$$NU_{-a} \subset BNU_a. \tag{9}$$

Now to show that $G = BNB_0^+$, it is enough to show that $BNB_0^+ g \subset BNB_0^+$ for all g in a set of generators of G. Hence it is enough to show that $BNB_0^+ U_{-a} \subset BNB_0^+$ for all $a \in \Pi_0$. But

$$\begin{aligned} BNB_0^+ U_{-a} &= BNU_a U^a U_{-a} = BNU_a U_{-a} U^a \\ &\subset BNU_a v_0^{-1}(s_a) U_a U^a \quad \text{by (2.6.6)} \\ &= BNv_0^{-1}(s_a) U_{-a} U^+ \\ &= BNU_{-a} U^+ \\ &\subset BNU_a U^+ \quad \text{by (9) above} \\ &= BNU^+ = BNB_0^+. \end{aligned}$$

□

Let Λ^+ (resp. Λ^{++}) be the set of all translations $t \in \Lambda$ such that $t(0) \in C_0^\perp$ (resp. $t(0) \in \overline{C_0}$). Let $Z^+ = v^{-1}(\Lambda^+), Z^{++} = v^{-1}(\Lambda^{++})$, so that Z^+ and Z^{++} are semi-groups contained in Z[12].

Also let K denote the maximal parahoric subgroup $G_{(0)}$.

Theorem 2.6.11

(1) *(Cartan decomposition)* $G = KZ^{++}K$, *and the mapping* $t \mapsto K \cdot v^{-1}(t) \cdot K$ *is a bijection of* Λ^{++} *onto* $K \backslash G/K$.

(2) *(Iwasawa decomposition)* $G = KZU^-$, *and the mapping* $t \mapsto K \cdot v^{-1}(t) \cdot U^-$ *is a bijection of* Λ *onto* $K \backslash G/U^-$.

(3) *If* $z_1 \in Z^{++}$ *and* $z_2 \in Z$ *are such that* $Kz_1K \cap Kz_2U^- \neq \emptyset$, *then* $z_1 z_2^{-1} \in Z^+$.

(4) *If* $z \in Z^{++}$, *then* $KzK \cap KzU^- = Kz$.

Proof As in §2.4 we shall identify N/H with W by means of the isomorphism induced by v. Thus KtK means $K \cdot v^{-1}(t) \cdot K$, and so on. In particular, $K = BW_0B$ by (2.3.2).

(1) If $w \in W$, then $KwK = BW_0wW_0B$ by (2.6.8) and (2.3.6). Now W is the disjoint union of the double cosets W_0tW_0 as t runs through Λ^{++}. Hence $G = BWB$ is the disjoint union of the cosets KtK with $t \in \Lambda^{++}$.

(2) We shall prove (2) with U^- replaced by U^+: this merely amounts to replacing the chamber C_0 by its opposite $-C_0$. From (2.6.10) we have

$$G = BWB_0^+ = BW_0B_0^+ \subset BW_0B \cdot B_0^+ = KB_0^+ = KZU^+.$$

To show that Kt_1U^+ and Kt_2U^+ are disjoint if $t_1 \neq t_2$, it is enough to show that $K \cap B_0^+ \subset HU^+$. So let $z_0u^+ \in K$, where $z_0 \in Z$ and $u^+ \in U^+$.

[12] Recall that C_0 is the chosen alcove which is a fundamental domain of W_0, and $C_0^\perp = \{x \in A : \langle x, y\rangle \geq 0, \forall y \in \overline{C_0}\}$. There is a mistake in the first edition which interchanges Λ^+ and Λ^{++}.

By (2.2.1) there exists $t \in \Lambda$ such that $u^+ \in U_{(tC_0)} = zU_{(C_0)}z^{-1} \subset zBz^{-1}$, where $z \in \nu^{-1}(t)$. Hence $z_0u^+ \in z_0zBZ^{-1} \cap K$, hence z_0zB meets Kz: or, putting $t_0 = \nu(z_0) \in \Lambda$, t_0tB meets $Kt = BW_0Bt$, and hence t_0tB meets $BwBt$ for some $w \in W_0$. Consequently $Bt_0tB \subset BwBtB$.

Now by expressing $w \in W_0$ as a reduced word in the generators s_a $(a \in \Pi_0)$, it follows easily from the axiom (BN3) that $BwBtB \subset BW_0tB$. Hence $Bt_0tB \subset BW_0tB$ and therefore $t_0t \in W_0t$ by (2.3.1). Hence $t_0 \in W_0 \cap \Lambda$, so that $t_0 = 1$ and $z_0 \in H$, as required.

The proofs of (3) and (4) are best expressed in the language of buildings §2.4.

(3) Suppose that $t_1 \in \Lambda^{++}$ and $t_2 \in \Lambda$ are such that Kt_1K meets Kt_2U^-. Then Kt_1^{-1} contains an element belonging to $u^-t_2^{-1}K$ for some $u^- \in U^-$. As above, by (2.2.1) we have $u^- \in U_{(-tC_0)} \subset G_{(C')}$ for some chamber C'. Hence Kt_1^{-1} meets $G_{(C')}t_2^{-1}K$. Let $x_1 = t_1(0)$ and $C'' = t_1(C')$. Then $G_{(x_1)} = t_1Kt_1^{-1}$ meets $t_1G_{(C')}t_2^{-1}K = G_{(C'')}t_1t_2^{-1}K$. Let $t_0 = t_1t_2^{-1}$ and put $x_0 = t_0(0)$.

Let $\mathcal{B}$ be the building associated with (B, N) as in §2.4. Identify the affine space A with the apartment $\mathcal{A}_0$ by means of the bijection j defined in (2.4.1). Let ρ be the retraction of $\mathcal{B}$ onto $\mathcal{A}_0$ with centre C'' (2.4.6) and let $g \in G_{(x_1)} \cap G_{(C'')}t_0K$. Since $g \in G_{(x_1)}$, the image under g of the segment $[0, x_1] \subset \mathcal{A}_0$ is the segment $[g(0), x_1]$. Since $g \in pt_0K$ for some $p \in G_{(C'')}$, we have

$$g(0) = pt_0(0) = p(x_0) \in p\mathcal{A}_0$$

and therefore $\rho(g(0)) = x_0$, $\rho(x_1) = x_1$ (because $x_1 \in \mathcal{A}_0$). Hence the image of the segment $[g(0), x_1]$ under the retraction ρ is a polygonal line γ in $\mathcal{A}_0$ joining x_0 to x_1. Consequently there exist real numbers λ_i with $0 = \lambda_0 < \lambda_1 < \cdots < \lambda_n = 1$ such that, if $y_i = \rho(g(\lambda_i x_i))$, then γ is the union of the segments $[y_{i-1}, y_i]$ $(1 \le i \le n)$. From (2.4.2) and the definition of ρ it follows that there exists elements $w_i \in W$ such that

$$[y_{i-1}, y_i] = w_i[\lambda_{i-1}x_1, \lambda_i x_1].$$

Writing $w_i = w_{i,0}\tau_i$ with $w_{i,0} \in W_0$ and $\tau_i \in \Lambda$, we have

$$y_i - y_{i-1} = (\lambda_i - \lambda_{i-1})w_{i,0}(x_1).$$

Now $x_1 \in \overline{C_0}$, hence $x_1 - wx_1 \in C_0^\perp$ for all $w \in W_0$. So we have

$$y_i - y_{i-1} = (\lambda_i - \lambda_{i-1})x_1 - y_i'$$

with $y_i' \in C_0^\perp$. Summing over i, we get

$$x_1 - x_0 = y_n - y_0 = x_1 - \sum y_i'$$

and therefore $x_0 = \sum y_i' \in C_0^\perp$. Since $x_0 = t_1t_2^{-1}(0)$, we have proved that $t_1t_2^{-1} \in \Lambda^+$, as required.

(4) Let $t \in \Lambda^{++}$ and let $g \in KtK \cap KtU^{-}$. Then by (2.2.1) as before we have $g^{-1} \in G_{(C_0')}t^{-1}K \cap Kt^{-1}K$ for some translate C_0' of the cone $-C_0$, and hence $g_1 = g^{-1}t \in G_{(C_0')}G_{(x)} \cap G_{(0)}G_{(x)}$, where $x = t^{-1}(0)$. Now $x \in -\overline{C_0}$, and $-\overline{C_0}$ intersects C_0'. Hence there exists $y \in C_0'$ such that x lies in the segment $[0y]$.

Let $g_1(x) = x' \in \mathcal{B}$. Since $g_1 \in pG_{(x)}$ for some $p \in G_{(0)} = K$, we have $x' = p(x)$ and $p(0) = 0$. Hence

$$d(0, x') = d(p(0), p(x)) = d(0, x).$$

Next, since $g_1 \in p'G_{(x)}$ for some $p' \in G_{(C_0')}$, we have $x' = p'(x)$ and $y = p'(y)$. Consequently

$$d(x', y) = d(p'(x), p'(y)) = d(x, y).$$

It follows that $w' \in [0y]$ by the uniqueness of geodesics in $\mathcal{B}$ (2.4.9) and hence that $x' = x$. Consequently g_1 fixes x, i.e. normalizes $G_{(x)}$ and therefore (2.3.6) $g_1 \in G_{(x)} = t^{-1}Kt$. Hence $g \in Kt$. □

2.7 Groups of p-adic type

Let G be an "abstract" group with affine root structure $(N, \nu, (U_\alpha)_{\alpha \in \Sigma})$ of type Σ_0. Now suppose in addition that G carries a Hausdorff topology compatible with its group structure such that

(VIII) *N and the U_α are closed subgroups of G;*
(IX) *B is a compact open subgroup of G.*

A Hausdorff topological group carrying an affine root structure satisfying axioms (I)–(IX) inclusive will be called a *simply connected group of p-adic type*. If k is a non-Archimedean local field (i.e. a locally compact field which is neither discrete nor connected) then k is complete with respect to a discrete valuation, and the residue field is finite. The universal Chevalley group $G(\Sigma_0, k)$, and more generally the group of k-rational points of a simply-connected simple linear algebraic group defined over k, then inherits a topology from k which satisfies (VIII) and (IX).

The two new axioms have the following consequences:

2.7.1 *H is open in N and compact.*

Proof $H = B \cap N$ by (2.6.8), so that H is an open subgroup of N, hence closed in N, hence closed in G. Since H is contained in B it follows that H is compact. □

2.7.2 *A closed subset M of G is compact if and only if M meets only finitely many cosets BwB.*

Proof The cosets BwB are mutually disjoint, open and compact. The assertion follows immediately from this. □

Comparing (2.7.2) with (2.4.13), we see that a subset M of G is bounded in the sense of §2.4 if and only if M has compact closure. Hence, from (2.4.15):

2.7.3 *Every compact subgroup of G is contained in a maximal compact subgroup. The maximal compact subgroups are precisely the maximal parahoric subgroups, and form $(\ell + 1)$ conjugacy classes. All parahoric subgroups are open and compact. G itself is locally compact, but not compact.*

Proof The last two assertions follow from the facts that a parahoric subgroup is a finite union of double cosets BwB, and that G itself is not. □

2.7.4 *For each $\alpha \in \Sigma$, the index $(U_\alpha : U_{\alpha+1})$ is finite, and U_α is compact.*

Proof There exists a chamber C' on which α is positive and for which h_α is one of the walls. Transforming by an element of W, we may assume that $C' = C$, i.e. that $\alpha \in \Pi$.

Since B is open and compact, $B/(B \cap s_\alpha B s_\alpha)$ is compact and discrete, hence *finite*. But

$$\begin{aligned} B/(B \cap s_\alpha B s_\alpha) &\cong (s_\alpha B s_\alpha \cdot B)/s_\alpha B s_\alpha \\ &= (s_\alpha B s_\alpha \cdot U_\alpha)/s_\alpha B s_\alpha \quad \text{by (2.6.7)} \\ &\cong U_\alpha/(U_\alpha \cap s_\alpha B s_\alpha) \\ &= U_\alpha/U_{\alpha+1} \end{aligned}$$

because $U_\alpha \cap s_\alpha B s_\alpha = U_\alpha \cap G_{(s_\alpha C)} = U_{\alpha+1}$ by (2.6.5). Hence $(U_\alpha : U_{\alpha+1})$ is finite. Also U_α is a closed subgroup of B, hence is compact. □

2.7.5

(1) *For each $a \in \Sigma_0$, the group $U_{(a)}$ is closed in G, and for each affine root α with gradient a, U_α is an open compact subgroup of $U_{(a)}$.*
(2) *U^+ is a closed subgroup of G and $U_{(C_0)}$ is an open compact subgroup of U^+.*

Proof By (2.6.5) $U_{(a)} \cap B$ is of the form U_α, where α is some affine root with gradient a. Hence this U_α is open in $U_{(a)}$. Conjugating U_α by elements of Z, each U_α is open in $U_{(a)}$. Since U_α is compact (2.7.4) it follows that $U_{(a)}$ is locally compact and therefore a closed subgroup of G. □

2.7.6 *$B_0^+ = ZU^+$ is locally compact.*

Proof Z is locally compact, because it has H as a compact open subgroup by (2.7.1). Also U^+ is locally compact by (2.7.5), hence so is B_0^+. □

2.7.7 Example Consider again the group $G = \mathrm{SL}_{\ell+1}(k)$ (2.5.1) where k is now a non-Archimedean local field. Let $\mathfrak{o}_k$ be the ring of integers of k, let $\mathfrak{p}$ be the maximal ideal of $\mathfrak{o}_k$, and $\kappa = \mathfrak{o}_k/\mathfrak{p}_k$ the (finite) residue field of k. $\mathfrak{o}_k$ is a compact open subring of k.

G is closed subset of $k^{(\ell+1)^2}$, hence is locally compact (and totally disconnected). The maximal compact subgroup K may be taken to be $\mathrm{SL}_{\ell+1}(\mathfrak{o}_k)$. Let $\overline{G} = \mathrm{SL}_{\ell+1}(\kappa)$, then the projection of $\mathfrak{o}_k$ onto κ induces an epimorphism $\pi\colon K \to \overline{G}$. Let $\overline{B}$ be the group of upper triangular matrices in $\overline{G}$, then $B = \pi^{-1}(\overline{B})$ is an Iwahori subgroup of G. B_0^+ is the group of upper triangular matrices in G, and Z is the diagonal matrices.

2.8 Notes on Chapter II

For proofs of the assertions in §2.1–§2.3, see Bourbaki [1]. The rest of the chapter is due to Bruhat and Tits [BT1, BT2].

2.9 Additions to Chapter II

2.9.1 Root datum and extended affine Weyl group

Definition A *root datum* is a quadruple $\mathcal{R} = (X, Y, \Sigma_0, \Sigma_0^\vee)$ such that X, Y are free abelian groups of finite rank together with a perfect pairing $\langle\ ,\ \rangle\colon X \times Y \to \mathbb{Z}$, $\Sigma_0 \subset X\backslash\{0\}$, $\Sigma_0^\vee \subset Y\backslash\{0\}$ are finite sets of roots and coroots respectively, in bijection $a \leftrightarrow a^\vee$ such that $\langle a, a^\vee\rangle = 2$, for every $a \in \Sigma_0$, the reflection $s_a : X \to X$, $s_a(x) := x - \langle x, a^\vee\rangle$ stabilizes Σ_0, and, for every $a \in \Sigma_0^\vee$, the reflection $s_{a^\vee} : Y \to Y$, $s_{a^\vee}(y) := y - \langle a, y\rangle$ stabilizes $\Sigma_0^\vee$.

The Weyl group W_0 of Σ_0, which is isomorphic to the Weyl group of $\Sigma_0^\vee$ via the map $a \mapsto a^\vee$, is called the Weyl group of $\mathcal{R}$. The root datum $\mathcal{R}$ is said to be irreducible if the root system Σ_0 is.

We set

$$V := \mathbb{Z}\Sigma_0 \otimes_{\mathbb{Z}} \mathbb{R} \quad \text{and} \quad V^* := \mathbb{Z}\Sigma_0^\vee \otimes_{\mathbb{Z}} \mathbb{R}$$

(the notation is consistent since $\mathbb{Z}\Sigma_0^\vee \otimes_{\mathbb{Z}} \mathbb{R}$ may be identified with the dual of $\mathbb{Z}\Sigma_0 \otimes_{\mathbb{Z}} \mathbb{R}$ via the pairing $\langle\ ,\ \rangle$). We denote by $\Sigma_0^+ \subset \Sigma_0$ the set of positive roots with respect to some linear order on V and by $\Pi_0 \subset \Sigma_0^+$ the corresponding set of simple roots. There are elements $\varpi_a \in V$ indexed by $a \in \Pi_0$ such that for every $b \in \Pi_0$

$$(\varpi_a, b^\vee) = \delta_{a,b} = \begin{cases} 1 & \text{if } a = b, \\ 0 & \text{otherwise.} \end{cases}$$

These elements form a basis of V and are called the *fundamental weights* of Σ_0. Dually, we can define the *fundamental coweights* $\varpi_a^\vee \in V^*$ by the property

$$\langle b, \varpi_a^\vee \rangle = \delta_{a,b}, \qquad \text{for every } b \in \Pi_0.$$

We consider respectively the *root lattice* and the *coroot lattice* of Σ_0:

$$Q := \mathbb{Z}\Sigma_0 = \bigoplus_{a \in \Pi_0} \mathbb{Z}a \subset V \quad \text{and} \quad Q^\vee := \mathbb{Z}\Sigma_0^\vee = \bigoplus_{a \in \Pi_0} \mathbb{Z}a^\vee \subset V^*. \tag{2.9.1}$$

Further, we have the respective *weight lattice* and *coweight lattice*:

$$P := \{\lambda \in V : \langle \lambda, a^\vee \rangle \in \mathbb{Z} \ \text{ for all } a \in \Sigma_0\} = \bigoplus_{a \in \Sigma_0} \mathbb{Z}\varpi_a \subset V, \tag{2.9.2}$$

$$\Lambda = P^\vee := \bigoplus_{a \in \Sigma_0} \mathbb{Z}\varpi_a^\vee \subset V^*. \tag{2.9.3}$$

We have

$$P^\vee \subset Y \subset Q^\vee.$$

For $\lambda, \mu \in P$, we write

$$\mu \leq \lambda \quad \text{if } \lambda - \mu = \sum_{i=1}^{\ell} n_i a_i \text{ with } n_i \in \mathbb{Z}_{\geq 0}. \tag{2.9.4}$$

It defines a partial order on P.

Definition The *extended affine Weyl group* W^{e} associated with $\mathcal{R}$ is

$$W^{\mathrm{e}} := W_0 \ltimes Y.$$

Remark The affine Weyl group $W = W^0 \ltimes P^\vee$ is a normal subgroup of W^{e}, and $W^{\mathrm{e}}/W \simeq Y/P^\vee$.

We set

$$P^+ := \{\lambda \in V : \langle \lambda, \alpha^\vee \rangle \in \mathbb{Z}_{\geq 0}, \forall \alpha \in \Sigma_0^+\} \tag{2.9.5}$$

and

$$P^{++} := \{\lambda \in V : \langle \lambda, \alpha^\vee \rangle \in \mathbb{Z}_{>0}, \forall \alpha \in \Sigma_0^+\}. \tag{2.9.6}$$

The elements of P^{++} are called *dominant weights*, and P^{++} constitutes a fundamental domain for P with respect to the action of W_0, in the sense that for any $\lambda \in P$ the orbit $P_{W_0}(\lambda) := \{w(\lambda) : w \in W_0\}$ intersects P^{++} precisely once. We say that λ is *antidominant* if $-\lambda$ is dominant. We define the *dominant coweights* and *antidominant coweights* analogously. For $\lambda \in P$, we denote by $W_0 \cdot \lambda$ the W_0-orbit of λ and by $W_{0,\lambda}$ the stabilizer of λ in W_0.

A non-zero dominant weight λ is called *minuscule* if $\langle \lambda, \alpha^\vee \rangle \leq 1$ for all $\alpha \in R^+$ and it is called *quasi-minuscule* if $\langle \lambda, \alpha^\vee \rangle \leq 1$ for all $\alpha \in \Sigma_0^+ \backslash \{\sigma\}$. The number of minuscule weights is one less than the *connection index* $|P/Q|$, which means that there are no minuscule weights if and only if the root lattice Q fills the whole weight lattice P. A quasi-minuscule weight, on the other hand, always exists and it is moreover unique: it is given by the dominant weight λ such that $\lambda^\vee$ is the maximal root of the dual root system $\Sigma_0^\vee$.

The collection of affine hyperplanes

$$h_{a+m} := \{v \in V : \alpha^\vee(v) = m\} \quad a \in \Sigma_0, m \in \mathbb{Z} \tag{2.9.7}$$

is W_0-stable. Let $\mathbb{H}$ denote the union of the hyperplanes h_{a+m}, with $a \in \Sigma_0$ and $m \in \mathbb{Z}$. The closures of the open connected components of $V \backslash \mathbb{H}$ are called *alcoves*. The length of an element w of W^e extends the length on W (defined in §2.1), i.e.,

$$\ell(w) := |\{h_\alpha : \alpha \in \Sigma \text{ and } h_\alpha \text{ separates } C \text{ and } w^{-1}C\}|, \tag{2.9.8}$$

where Σ is the set of affine roots attached to Σ_0 (as in §2.2) and C any alcove (e.g. the fundamental one). The group W^e permutes the alcoves. The stabilizer of C is isomorphic to the group Ω. Hence Ω consists of elements in W^e of length 0, and s_a permutes simply transitively the alcoves, for every $a \in \Sigma_0$.

Definition A *Coxeter group* W' is a group generated by a set S of elements of order 2, which has a presentation

$$W' = \langle S : (ss')^{m(s,s')} = 1 \text{ for all } s, s' \in S \rangle,$$

where $m(s, s') \in \mathbb{Z}_{\geq 1}$ is the order of ss' in W'. The equalities $s^2 = 1$ (for $s \in S$) are called the *quadratic relations*, while the equalities $(ss')^{m(s,s')} = 1$, or equivalently

$$\underbrace{ss'ss'\cdots}_{m(s,s')\text{ terms}} = \underbrace{s'ss's\cdots}_{m(s,s')\text{ terms}} \tag{2.9.9}$$

are called the *braid relations*.

We keep the notation of §2.9.1. We set $S_0 := \{s_a : a \in \Pi_0\}$ and $S_{\text{aff}} := \{s_a : a \in \Pi\}$. Let $\Sigma_{0,\max}^\vee$ be the set of maximal elements of $\Sigma_0^\vee$, with respect to the base $\Pi_0^\vee := \{a^\vee : a \in \Pi_0\}$. It contains one element for every irreducible component of $\Sigma_0^\vee$. For $a^\vee \in \Sigma_{0,\max}^\vee$ we define

$$s'_a : X \to X, \; s'_a(x) := x + a - \langle x, a^\vee \rangle a. \tag{2.9.10}$$

This is the reflection of $X \otimes_{\mathbb{Z}} \mathbb{R}$ in the hyperplane h_{a+1}. Then

$$S_{\text{aff}} = S_0 \cup \{s'_a : \alpha^\vee \in \Sigma_{0,\max}^\vee\}. \tag{2.9.11}$$

The groups (W_0, S_0) and (W, S_{aff}) are both Coxeter groups.

2.9.2 Affine Hecke algebras

Spherical functions (on p-adic symmetric spaces) – also referred to as generalized Hall–Littlewood polynomials associated with root systems – are intimately connected with the theory of affine Hecke algebras. The classical theory of Hall–Littlewood polynomials and the Kostka–Foulkes polynomials appears in [Mac2].

Definition Let (W', S) be a Coxeter group, equipped with a function $\mathbf{q}\colon s \mapsto q_s$ from S to $\mathbb{C}$ satisfying

$$q_s = q_{s'} \quad \text{if } s \text{ and } s' \text{ are conjugate under } W'. \tag{2.9.12}$$

There is a unique algebra structure $\mathcal{H}(W', \mathbf{q})$ on the $\mathbb{C}$-vector space spanned by elements T_w $(w \in W')$ such that

(1) $T_1 = 1$,
(2) $(T_s - q_s)(T_s + 1) = 0$ for any $s \in S$ (quadratic relations),
(3) $T_s T_{s'} T_s \cdots = T_{s'} T_s T_{s'} \cdots$ where both sides have $m_{s,s'}$ elements (braid relations),
(4) $T_{ww'} = T_w T_{w'}$ if $\ell(ww') = \ell(w) + \ell(w')$.

The algebra $\mathcal{H}(W', \mathbf{q})$ is called the *Hecke–Iwahori algebra* of $(W', \mathbf{q})$.

Example Let I be an Iwahori subgroup of a p-adic group G. The algebra $\mathcal{H}(G, I)$ of locally constant compactly supported bi-I invariant functions with the multiplication given by convolution is an Hecke–Iwahori algebra. A list of its properties can be found in [HKP]. In particular, $\mathbb{C}[\Lambda]$ is a commutative algebra of $\mathcal{H}(G, I)$, and the finite-dimensional Hecke algebra $\mathcal{H}(K, I)$ is a subalgebra of $\mathcal{H}(G, I)$.

We fix $q \in \mathbb{R}_{>1}$ and let $\mathbf{q}\colon S_0 \to \mathbb{C}$ be the function $s \mapsto q_s$, and let $\mathcal{H}(W_0, \mathbf{q})$ be the Iwahori–Hecke algebra of W_0.

Definition Let $\{\theta_\lambda : \lambda \in \Lambda\}$ denote the standard basis of $\mathbb{C}[\Lambda]$. The *affine Hecke algebra* $\mathcal{H}(W, \mathbf{q})$ (resp. extended affine Hecke algebra $\mathcal{H}(W^{\mathrm{e}}, \mathbf{q})$) is the vector space

$$\mathbb{C}[Y] \otimes_{\mathbb{C}} \mathcal{H}(W_0, \mathbf{q}) \quad (\text{resp. } \mathbb{C}[\Lambda] \otimes_{\mathbb{C}} \mathcal{H}(W_0, \mathbf{q}))$$

such that $\mathbb{C}[Y]$ (resp. $\mathbb{C}[\Lambda]$) and $\mathcal{H}(W_0, \mathbf{q})$ are embedded as subalgebras, and for $a \in \Pi_0$ and $x \in X$:

$$\theta_\lambda T_{s_a} - T_{s_a} \theta_{s_a(\lambda)} = \left((q_{s_a} - 1) + \theta_{-a} \left(q_{s_a} q_{s'_a} - q_{s_a} q_{s'_a}^{-1} \right) \right) \frac{\theta_x - \theta_{s_a(x)}}{\theta_0 - \theta_{-2a}}. \tag{2.9.13}$$

Note that when $a^\vee \notin 2P^\vee$, the cross relation simplifies to

$$\theta_\lambda T_{s_a} - T_{s_a} \theta_{s_a(\lambda)} = (q_{s_a} - 1) \frac{\theta_\lambda - \theta_{s_a(\lambda)}}{\theta_0 - \theta_{-a}}.$$

The centre of $\mathcal{H}(W^{\mathrm{e}}, \mathbf{q})$ is the ring

$$\mathrm{Z}(\mathcal{H}(W^{\mathrm{e}}, \mathbf{q})) := \{ f \in \mathbb{C}[X] : w \cdot f = w \text{ for all } w \in W_0 \} = \mathbb{C}[X]^{W_0} \tag{2.9.14}$$

of symmetric functions in $\mathbb{C}[X]$ (see [Lus2, Proposition 3.11] or [NR, Theorem 1.33]). Similarly, the centre of $\mathcal{H}(W, \mathbf{q})$ is $\mathbb{C}[\Lambda]^{W_0}$.

2.9.3 The simplicial approach to buildings

Definition A *chamber complex* is a simplicial complex $\mathcal{B}$ such that all simplices of $\mathcal{B}$ are contained in a maximal simplex and such that for every pair x, y of maximal simplices of $\mathcal{B}$ there is a chain $x = x_0, x_1, \ldots, x_n = y$ of adjacent maximal simplices in $\mathcal{B}$. Its maximal simplices are called *chambers*.

A chamber complex is called a *thin chamber complex* if each facet of a chamber is the facet of exactly two chambers, and it is called *thick* if each facet of a chamber is the facet of at least three chambers.

An important example of a thin chamber complex is the *Coxeter complex* $\Sigma(W, S)$ associated with a Coxeter system (W, S). Its simplices are indexed by the cosets W/I, where $I \subset S$ is a proper subset. The face relation is determined by opposite inclusion. In particular, the singletons $\{w\}$ represent the maximal simplices of $\Sigma(W, S)$.

Definition A *system of apartments* in a chamber complex $\mathcal{B}$ is a set $\mathfrak{A}$ of thin chamber subcomplexes such that

- for each pair of simplices x, y of X there is an $\mathcal{A} \in \mathfrak{A}$ such that $x, y \in \mathcal{A}$, and
- if $\mathcal{A}, \mathcal{A}' \in \mathfrak{A}$ contain a common chamber C and a simplex x, then there is an isomorphism of chamber complexes $\mathcal{A} \to \mathcal{A}'$ that fixes C and x point-wise.

A maximal element among non-chamber facets is called a *panel*; it is an open subset of a wall (called its *support*). A (thick) *building* is a (thick) chamber complex that admits a system of apartments.

One can prove that for a thick building $\mathcal{B}$ there is a unique maximal apartment system. So if $\mathcal{B}$ is thick, as we assume from now on, it makes sense to speak of an apartment of $\mathcal{B}$ without reference to a specific system of apartments. There is a Coxeter system (W, S) such that every apartment of $\mathcal{B}$ is chamber isomorphic with $\Sigma(W, S)$. The choice of a fundamental chamber C in $\mathcal{B}$ and a compatible labeling of its facets by elements of S makes these isomorphisms unique and introduces a labeling of all facets of chambers in $\mathcal{B}$. The Coxeter system (W, S) is called the *type* of $\mathcal{B}$.

An *irreducible* (resp. *affine*) building is a building whose Coxeter system is irreducible (resp. affine, as we assume throughout). The rank of $\mathcal{B}$ is by definition the rank of (W, S), which is equal to the number of facets of any chamber in $\mathcal{B}$. We denote by $V(\mathcal{B})$ the set of vertices of $\mathcal{B}$.

The thickness of $\mathcal{B}$ is the tuple $(d_s)_{s \in S}$, where $d_s - 1$ is the cardinality of the set of chambers containing the fundamental chamber's facet labeled by s (so that a building is thick if and only if $d_s \geq 2$ for all $s \in S$). The building is called *locally finite* if all d_s, $s \in S$ are finite.

Given a locally finite, thick building $\mathcal{B}$, its group of simplicial automorphisms $\mathrm{Aut}(\mathcal{B})$ carries the topology of point-wise convergence on simplices, which turns $\mathrm{Aut}(\mathcal{B})$ into a totally disconnected, locally compact group. A subgroup G of $\mathrm{Aut}(\mathcal{B})$ is called *type preserving* if it leaves invariant any labeling of facets of chambers in X obtained from the choice of a labeled fundamental chamber. A subgroup $G \subset \mathrm{Aut}(\mathcal{B})$ is called strongly transitive if it acts transitively on pairs $(\mathcal{A}, C)$, where $\mathcal{A}$ is an apartment of $\mathcal{B}$ and C is a chamber in $\mathcal{A}$.

Given a closed, strongly transitive subgroup $G \subset \mathrm{Aut}(\mathcal{B})$, an *Iwahori subgroup* of G, denoted below by I, is the stabiliser of a chamber in $\mathcal{B}$. Every Iwahori subgroup is compact and open and it is unique up to conjugation by elements from G. Iwahori subgroups play the role of the group B in (B, N)-pairs, which can be used to construct buildings from groups. If (W, S) is the type of $\mathcal{B}$, then the Bruhat decomposition of a closed, strongly transitive and type preserving subgroup G of $\mathrm{Aut}(\mathcal{B})$ provides a bijection

$$W \longrightarrow I\backslash G/I \tag{2.9.15}$$

assigning to $w \in W$ its Bruhat cell IwI. Note that this identification uses the fact that W can be viewed as the group of type preserving automorphisms of an apartment of X.

Bruhat–Tits buildings (also known as *Euclidean* buildings, or *affine* buildings) are metric spaces of non-positive curvature that provide p-adic analogues of Riemannian symmetric spaces. They have several applications in group theory that are of geometric and topological interest. The simplest buildings are trees, and can be used to prove Ihara's Theorem that every torsion-free discrete subgroup of $\mathrm{SL}_2(\mathbb{Q}_p)$ is free [Ih]. The link of a vertex in a Bruhat–Tits building is a Tits building. (Also, a Bruhat–Tits building can be compactified by adding a sphere at ∞, and this boundary sphere has the structure of a Tits building.)

There are several good references on buildings from different interesting viewpoints: Tits' original lecture notes from 1974 [Ti1] can still serve as a very good introduction to the subject, [Gar] considers the case of classical groups, more complete references are [Ron] and [AB] for general buildings, [Ti1] and [Wei1] for Tits buildings, [Ti2], [BT1, BT2, BT3], [Wei2] and [KP] for Bruhat–Tits buildings.

2.9.4 Parahoric subgroups and Moy–Prasad filtrations

If as an algebraic group over the algebraic closure of k the group G is simply connected, then the maximal compact subgroups of G are parahoric subgroups. Conversely, a parahoric subgroup is a maximal subgroup of G if and only if it is maximal among parahoric subgroups. In fact, the maximal compact subgroups of G are the parahoric subgroups whose fixators in the Bruhat–Tits buildings are 0-dimensional facets. If G is not simply connected, then the maximal compact subgroups are not parahoric subgroups in general: they may be bigger. For example, the normalizer of the standard Iwahori subgroup of $\mathrm{PGL}_n(k)$ is a maximal compact subgroup which is not a parahoric subgroup. Compact subgroups are not necessarily stabilizers of vertices. Stabilizers of vertices are compact modulo centre, hence if G is semisimple they are compact. When G is both simply connected and semisimple, the stabilizers of vertices coincide with the parahoric subgroups of G.

Let k_s be a fixed separable closure of k and $\Gamma_s := \mathrm{Gal}(k_s/k)$. We denote by k_{nr} the maximal unramified extension of k contained in k_s. Let G be a connected reductive k-group. We define a homomorphism

$$\mathrm{val}_G : G \to \mathrm{Hom}_{\mathbb{Z}}(X^*(G)^{\Gamma_s}, \mathbb{Z}), \quad g \mapsto (\chi \mapsto -\mathrm{val}_k(\chi(g))), \tag{2.9.16}$$

and denote by G_1 the kernel of val_G. Let $\mathfrak{G}$ be a smooth $\mathfrak{o}$-group scheme, and let $\mathfrak{G}^\circ$ denote the largest subgroup scheme of $\mathfrak{G}$ with connected fibers. By [SGA3, Exp.VIB, Theorem 3.10], $\mathfrak{G}^\circ$ exists and is an open subscheme of $\mathfrak{G}$. By [PY, §3.5], if $\mathfrak{G}$ is affine, then so is $\mathfrak{G}^\circ$.

Given a non-empty bounded subset E of an apartment $\mathcal{A}$ of $\mathcal{B}$, we denote by $(G_1)_E$ the subgroup of G_1 consisting of elements that fix E pointwise. There is a smooth affine $\mathfrak{o}_k$-group scheme $\widetilde{\mathfrak{G}}_E$, with generic fiber G, such that $\widetilde{\mathfrak{G}}_E(\mathfrak{o}) = (G_1)_E$. The three following statements are proved in [KP, Theorem 4.2.15]. If x is vertex of $\mathcal{B}(G)$, then the group $(G_1)_x$ is a maximal open and bounded subgroup of G. Every maximal bounded subgroup of G is of the form $(G_1)_x$ for a point x of $\mathcal{B}(G)$. If $G^\circ = G_1$ (this holds for example when G is semi-simple and simply connected) then x is a vertex.

Let $\mathbf{T}$ be a k-torus. The group $T := \mathbf{T}(k)$ has a unique maximal bounded subgroup T_b, which has the following description (see for instance [KP, Propositions 2.2.12 and 2.5.8]):

$$T_b = \{t \in T \,:\, \mathrm{val}(\chi(t)) = 0 \;\text{ for all } \chi \in X^*(\mathbf{T})\} =: T^1.$$

Let $T^0 \subset T$ denote the image of $\mathbf{T}(k')_b$ under the norm map $\mathbf{T}(k') \to \mathbf{T}(k)$ for any finite Galois extension k'/k splitting $\mathbf{T}$ and let $T^0 := \mathbf{T}(k')^0 \cap T$. The subgroup T^0 lies in T^1 as a subgroup of finite index, and the groups $\mathbf{T}(k')^0$ and T^0 are Iwahori subgroups of $\mathbf{T}(k')$ respectively T.

A k-torus $\mathbf{T}$ is called *induced* if the lattice $X^*(\mathbf{T})$, equivalently $X_*(\mathbf{T})$, has a $\mathbb{Z}$-basis that is invariant under the Galois group of the splitting extension of $\mathbf{T}$, and *weakly induced* if there exists a finite tamely ramified field extension k'/k such that $\mathbf{T}_{k'}$ is an induced torus.

In applications, the main examples of induced tori are when $\mathbf{T}$ is a maximally split maximal torus of a quasi-split connected reductive group G that either splits over a tame extension, or is simply connected, or adjoint. From now on, we will assume that T is weakly induced. The Moy–Prasad filtration of T is then defined by $T_0 := T^0$, and for $r > 0$:

$$T_r := \{t \in T_0 \,:\, \mathrm{val}(\chi(t) - 1) \geq r \;\text{ for all } \chi \in X^*(\mathbf{T})\}. \tag{2.9.17}$$

It gives a decreasing filtration of T^1 by bounded open subgroups T_r, which is separated.

Let $\mathbf{S}$ be a maximal k-split torus of $\mathbf{G}$, and let $\mathbf{Z}$ denote the centralizer of $\mathbf{S}$ in $\mathbf{G}$. If $\mathbf{G}$ is quasi-split, then $\mathbf{Z}$ is a maximal torus $\mathbf{T}$, and the filtration $\mathbf{Z}(k)_r$ is defined by (2.9.17). We will suppose for the moment that the filtration $\mathbf{Z}(k)_r$ has been defined also when $\mathbf{G}$ is not quasi-split.

The root subgroup U_a of G, for $a \in \Sigma_0$, admits the following filtration by compact open subgroups U_α, indexed by those $\alpha \in \Sigma$ with gradient a. Given a point $x \in A$, and $a \in \Sigma_0$, we construct a valuation φ_a^x of the root group U_a as follows. Given $u \in U_a^*$, there is a real-valued affine function ψ_a^u on A, with derivative a, such that the set of points of A fixed by u is the closed half-apartment $\{x \in A : \psi_a^u(x) \geq 0\}$ (see for instance [KP, Proposition 9.4.3]). For $u \in U_a$, we set $\varphi_a^x(u) := \psi_a^u(x)$. We set $\varphi_a^x(1) := \infty$.

Let $x \in A$. We set

$$U_{a,x,r} := (\varphi_a^x)^{-1}([r, \infty]). \tag{2.9.18}$$

Let $\widetilde{\Sigma}_0 := \Sigma_0 \cup \{0\}$. A function $f: \Sigma \to \mathbb{R}$ is called *concave* if $f(a+b) \leq f(a) + f(b)$ for any $a, \in \Sigma_0$ with $a + b \in \Sigma_0$. For any $r \in \mathbb{R}$, the constant function $f: \Sigma_0 \to \mathbb{R}$ taking the value r is concave.

Let $f: \Sigma_0 \to \mathbb{R}$ be a concave function. We define the following subgroup of G:

$$U_{a,x,f} := U_{a,x,f(a)} \cdot U_{2a,x,f(2a)}. \tag{2.9.19}$$

We denote by $U_{x,f}$ the subgroup of G generated by the $U_{a,x,f}$ for all $a \in \Sigma_0$, and set

$$G_{x,f} := U_{x,f} \cdot \mathbf{Z}(k)_{f(0)}. \tag{2.9.20}$$

Let F be the facet containing x. The function $f = 0$ is concave, and the subgroup $G_{x,0}$ coincides with the parahoric subgroup G_F of G. We define $G_{x,r}$ to be the (open bounded) subgroup of G generated by $\mathbf{Z}(k)_r$ and the $U_{a,x,r}$ for all $a \in \Sigma_0$. Note that, while the parahoric subgroup $G_{x,0}$ and its subgroup $G_{x,0+}$ depend only on the facet containing x, in general this is not true for $G_{x,r}$, which may well depend on the particular point x within that facet.

We come back to the definition of $\mathbf{Z}(k)_r$ in the case when G is not quasi-split. The building $\mathcal{B}(Z)$ of Z consists of a single point x. The group $\mathbf{Z}(k_{\mathrm{nr}})$ is quasi-split and the point $x \in \mathcal{B}(Z) \subset \mathcal{B}(\mathbf{Z}(k_{\mathrm{nr}}))$. By applying the construction above to the group $\mathbf{Z}(k_{\mathrm{nr}})$, we obtain the subgroups $\mathbf{Z}(k_{\mathrm{nr}})_{x,r}$. We set $\mathbf{Z}(k)_r := \mathbf{Z}(k_{\mathrm{nr}})_{x,r} \cap \mathbf{Z}(k)$.

We set

$$G_{x,r+} := \bigcup_{t>r} G_{x,t}. \tag{2.9.21}$$

When $r > 0$, the quotient $G_{x,r}/G_{x,r+}$ is abelian and can be identified with a vector space over the residual field κ of k.

The formalism of Moy and Prasad has been notably exploited in [AD] to sharpen and extend familiar harmonic analysis results for $\mathfrak{g}$, showing for instance that the G-orbits of the Moy–Prasad filtration lattices are asymptotic to the set of nilpotent elements in $\mathfrak{g}$, defining G-domains in terms of the filtration lattices and studying their properties.

The Berkovich space G^{an} attached to G is a very rich and interesting object. Bruhat–Tits buildings can be realized inside Berkovich spaces. In this way, Berkovich analytic geometry can be used to compactify buildings. An intrinsic description of Bruhat–Tits buildings in the framework of non-Archimedean analytic geometry was given in [RTW]. Berkovich (for G split), and later Rémy, Thuillier and Werner (for G arbitrary) showed that the reduced Bruhat–Tits building of G embeds canonically in G^{an}. In [May], Mayeux associated to every pair (x, r) formed by a special point x in the building and a positive rational number r an element $\theta(x, r)$ of G^{an} whose projective holomorphic envelope is an analytic group $G_{x,r}$. It is a generalization of the construction made in [RTW] in the case $r = 0$. The element $\theta(x, r)$ is defined to be the Shilov boundary of the image by the projection $(G \times_{\mathrm{Spec}(k)} \mathrm{Spec}(k'))^{\mathrm{an}} \to G^{\mathrm{an}}$, where k' is an extension of k, that depends on x and r, of the generic fiber of the formal completion of a congruence subscheme of a Demazure group scheme on the ring of integers of k along its special fiber.

2.9.5 Representations of p-adic groups

The theory of complex representations of p-adic groups has been initiated by Mautner in [Mau2, Mau3]. First general results have been obtained by Bruhat in [Bru2], who adopted Schwartz's theory of distributions as the proper language for studying harmonic analysis on p-adic groups. The next significant progress was due to Satake, who determined the spherical functions on reductive p-adic groups [14]. Later developments in the theory benefited a lot from the monumental work of Bruhat and Tits [BT1] on the internal structure of reductive p-adic groups, also known as the theory of affine buildings. At the same time, a general theory of harmonic analysis on p-adic groups was being built up by Harish-Chandra on the model of Lie groups. Harish-Chandra enunciated the *philosophy of cusp forms* and stated the Plancherel formula, both of which are tremendously influential. The theory of representations of p-adic reductive groups has nowadays attained a mature stage of development. A large class of supercuspidal representations have been constructed. The local Langlands correspondence is now established in many cases. On the other hand, many deep questions remain open.

Apart from studying representations of p-adic groups for their own sake, a great source of motivation stems from the realization of automorphic forms as representations of adelic groups, whose representations of p-adic groups are local components. In recent decades, the study of congruences between modular forms has had tremendous impact in Number Theory, leading in particular to a proof of Fermat's Last Theorem. This led to the study of congruences between automorphic forms and automorphic representations. As a local counterpart, one also has to explore representations of p-adic groups on C-vector spaces, where C is a field of positive characteristic, for example a finite field

Let G be a p-adic reductive group (that is, k is a non-Archimedean local field) and K a maximal compact subgroup of G.

Definition A representation $(\pi, \mathcal{V})$ of G, where $\mathcal{V}$ is a $\mathbb{C}$-vector space, is said to be *smooth* if for each $v \in \mathcal{V}$, the stabilizer of v in G is open.

Let $\mathfrak{R}(G)$ denote the category of all smooth complex representations of G. This is an abelian category admitting arbitrary coproducts. The algebra of endomorphisms of the identity functor of the category $\mathfrak{R}(G)$, which was introduced in [Be1], is called the *Bernstein centre* of G. We denote it by $\mathfrak{Z}(\mathfrak{R}(G))$. In the representation theory of reductive p-adic groups the algebra $\mathfrak{Z}(\mathfrak{R}(G))$ plays the role of the centre of the universal enveloping algebra in the representation theory of real reductive groups.

Smooth and compact inductions

Let H be a closed subgroup of G and let $(\sigma, \mathcal{V})$ be a smooth representation of H. We define the smooth representations of G:

$$\mathrm{Ind}_H^G(\sigma) := \{f \in C^\infty(G, V) \,:\, f(hg) = \sigma(h)f(g) \text{ for all } h \in H\}, \qquad (2.9.22)$$

$$\mathrm{c{-}Ind}_H^G(\sigma) := \left\{f \in \mathrm{Ind}_H^G(\sigma) \,:\, \text{the support of } f \text{ is compact modulo } H\right\}, \qquad (2.9.23)$$

where G acts by right translation.

A subgroup P of G is parabolic if and only if the variety G/P is proper, in which case it is projective (this variety is called a (partial) flag variety). The parabolic subgroups of G, as well as the parahoric subgroups of G, form a building, called the *Tits building* or *spherical building* of G. For instance, the statement that any two parabolic subgroups of G contain a common maximal k-split torus is analogous to the statement that any two parahoric subgroups of G contain the maximal bounded subgroup of a maximal k-split torus. When G is simply connected semisimple, the statement that a parabolic subgroup of G is any subgroup that contains a minimal parabolic subgroup is analogous to the statement that a parahoric subgroup is any bounded subgroup of G that contains an Iwahori subgroup.

Let P be a parabolic subgroup of G with unipotent radical U and let L be a Levi factor of P, so $P = LU$ and we have a canonical isomorphism $L \simeq P/U$ obtained by composing the injection $L \hookrightarrow P$ with quotient map $P \to P/U$. Then P is a closed subgroup of G with a compact quotient G/P. The inflation of a representation σ of L to P is the unique representation of P which restricts to σ on L and is trivial on U; we denote it by $\mathrm{infl}_L^P(\sigma)$. Then the *parabolic induction* functor is defined as the following composition

$$\mathrm{i}_{L,P}^G \colon \mathfrak{R}(L) \xrightarrow{\mathrm{inf}_L^P} \mathfrak{R}(P) \xrightarrow{\mathrm{Ind}_P^G} \mathfrak{R}(G). \qquad (2.9.24)$$

We denote by $\mathrm{I}_{L,P}^G$ the functor of normalised parabolic induction defined by

$$\mathrm{I}_{L,P}^G(\sigma) := \mathrm{i}_{L,P}^G(\delta^{1/2} \otimes \sigma), \quad \text{where } \delta(l) = |\det(\mathrm{Ad}(l)|_{\mathfrak{p}/\mathfrak{l}})|_k. \qquad (2.9.25)$$

An irreducible smooth representation π of G is said to be *supercuspidal* if it is not a subquotient of a proper parabolically induced representation.

Any smooth irreducible π of G occurs as an irreducible component of a parabolically induced representation $\mathrm{i}_{L,P}^{G}(\sigma)$, where P is a parabolic subgroup of G with Levi factor L and σ is a supercuspidal irreducible smooth representation of L.

The G-conjugacy class $(L,\sigma)_G$ of (L,σ) is uniquely determined and is called the *supercuspidal support* of π. It is denoted by $\mathrm{Sc}(\pi)$.

Definition An irreducible smooth representation π of G is *unipotent* if there exists an element x of the Bruhat–Tits building of G, a parahoric subgroup of G, and an irreducible representation $\widetilde{\tau}$ of the parahoric subgroup $G_{x,0}$, which is inflated from a *cuspidal unipotent* representation τ of the reductive quotient $\mathbb{G}_{x,0}$ (rational points of a connected reductive group over the residue field k_F of F) of $G_{x,0}$, such that the restriction of π to $G_{x,0}$ contains $\widetilde{\tau}$.

Example Every Iwahori-spherical irreducible representation of G is unipotent.

Remark The supercuspidal irreducible unipotent representations of G are the representations $\mathrm{c\text{-}Ind}_{\mathrm{N}_G(G_{x,0})}^{G}\widetilde{\tau}$ such that x is a vertex in the Bruhat–Tits building of G, and τ is an irreducible cuspidal unipotent of $\mathbb{G}_{x,0}$.

The Bernstein centre of G can be realized as the algebra of regular functions on the algebraic variety $\Omega(G)$, called the *Bernstein variety* of G, defined as

$$\Omega(G) := \{(L,\sigma)_G \; : \; L \text{ a Levi subgroup of } G,\ \sigma \text{ an irreducible supercuspidal of } L\}. \tag{2.9.26}$$

Here $(L,\sigma)_G$ denotes the G-conjugacy class of the pair (L,σ).

Let G^1 denote the subgroup of G generated by all compact subgroups. Then G^1 is an open normal subgroup with compact center, $\Lambda(G) := G/G^1$ is a lattice, and $G^1 \cdot Z_G$ is of finite index in G.

Example If $G = \mathrm{GL}_n(k)$, we have $G^1 = \left\{g \in G \; : \; \det(g) \in \mathfrak{o}_k^\times\right\}$, and $\Lambda(G) \simeq \mathbb{Z} = k^\times/\mathfrak{o}_k^\times$.

A character of G is said to be *unramified* if it is trivial on G^1. Let $\mathfrak{X}_{\mathrm{nr}}(G)$ be the group of unramified characters of G. Every $\chi \in \mathfrak{X}_{\mathrm{nr}}(G)$ is of the form $|\chi_1|_k^{s_1}\cdots|\chi_m|_k^{s_m}$, where $\chi_1, \ldots, \chi_m$ are k-rational algebraic characters of G (i.e. homomorphisms into $\mathbb{G}_m$), defined over k, and $s_1, \ldots, s_m \in \mathbb{C}$. The group $\mathfrak{X}_{\mathrm{nr}}(G)$ has a natural structure of a complex algebraic torus: if $\chi_1, \ldots, \chi_m$ form a basis for the group of algebraic characters modulo torsion and if $\chi = |\chi_1|_k^{s_1}\cdots|\chi_r|_k^{s_m}$, then the association $\chi \mapsto (q-s_1, \ldots, q-s_m) \in (\mathbb{C}^\times)^m$ defines the structure of a complex torus on $\mathfrak{X}_{\mathrm{nr}}(G)$.

Definition Let $P_0 = L_0U_0$ be a minimal parabolic subgroup of G with Levi factor L_0 and let $\chi \in \mathfrak{X}_{\mathrm{nr}}(L_0)$. The set of irreducible constituents of $\mathrm{I}_{L_0,P_0}^{G}(\chi)$ is called the *unramified principal series representation* of G induced by χ.

Let $(\sigma, \mathcal{W})$ be an irreducible supercuspidal smooth representation of L and let $\mathcal{O}$ denote the orbit of σ under the action of $\mathfrak{X}_{\mathrm{nr}}(L)$:

$$\mathcal{O} := \{\sigma \otimes \chi \, : \, \chi \in \mathfrak{X}_{\mathrm{nr}}(L)\} \,. \tag{2.9.27}$$

We write $\mathfrak{s} := (L, \mathcal{O})_G = [L, \sigma]_G$ for the G-conjugacy class of the pair $(L, \mathcal{O})$ and $\mathfrak{B}(G)$ for the set of such classes $\mathfrak{s}$. We set $\mathfrak{s}_L := (L, \mathcal{O})_L$.

Let $\mathfrak{R}^{\mathfrak{s}}(G)$ denote the full subcategory of $\mathfrak{R}(G)$ whose objects are the representations π such that every irreducible G-subquotient of π has its supercuspidal support in $\mathfrak{s}$. The categories $\mathfrak{R}^{\mathfrak{s}}(G)$ are indecomposable and split the full smooth category $\mathfrak{R}(G)$ into a direct product:

$$\mathfrak{R}(G) = \prod_{\mathfrak{s} \in \mathfrak{B}(G)} \mathfrak{R}^{\mathfrak{s}}(G) \tag{2.9.28}$$

(see [Be1], [Be2], [Ren] or [Ro2]). By the decomposition (2.9.28), we evidently have

$$\mathfrak{Z}(\mathfrak{R}(G)) = \prod_{\mathfrak{s} \in \mathfrak{B}(G)} \mathfrak{Z}(\mathfrak{R}^{\mathfrak{s}}(G)), \tag{2.9.29}$$

where $\mathfrak{Z}(\mathfrak{R}^{\mathfrak{s}}(G))$ is the centre of the category $\mathfrak{R}^{\mathfrak{s}}(G)$.

The category of unipotent representations of G is

$$\mathfrak{R}^{\mathrm{unip}}(G) := \prod_{\mathfrak{s} \in \mathfrak{B}^{\mathrm{unip}}(G)} \mathfrak{R}^{\mathfrak{s}}(G), \tag{2.9.30}$$

where $\mathfrak{B}^{\mathrm{unip}}(G)$ is the following subset of $\mathfrak{B}(G)$:

$$\{[L, \sigma]_G \, : \, \sigma \text{ a supercuspidal irreducible unipotent representation of } L\} \,.$$

Example We suppose that G is quasi-split. Let $\mathfrak{i} := [T, \chi]_G$, where $T \subset B$ is a maximal torus and χ is an unramified character of T. The centre of $\mathfrak{R}^{\mathfrak{i}}(G)$ is naturally isomorphic to $\mathbb{C}[\mathfrak{X}_{\mathrm{nr}}(T)]^{W_0}$ by mapping each element to the scalar by which it acts on $\mathrm{I}_{T,B}^G(\chi)$, for $\chi \in \mathfrak{X}_{\mathrm{nr}}(T)$.

Let $\mathrm{Irr}^{\mathfrak{s}}(G)$ denote the set of irreducible objects of the category $\mathfrak{R}^{\mathfrak{s}}(G)$. As a direct consequence of (2.9.28), we have

$$\mathrm{Irr}(G) = \prod_{\mathfrak{s} \in \mathfrak{B}(G)} \mathrm{Irr}^{\mathfrak{s}}(G). \tag{2.9.31}$$

As noticed in [Ro1, §1.2], the isomorphism class of $\Pi^{\mathfrak{s}_L} := \mathrm{c}\text{-}\mathrm{Ind}_{L^1}^{L}(\sigma_1)$ does not depend on the choice of σ_1. It was shown by Bernstein that

$$\Pi^{\mathfrak{s}} := \mathrm{I}_{L,P}^G(\Pi^{\mathfrak{s}_L}) \tag{2.9.32}$$

is a progenerator of $\mathfrak{R}^{\mathfrak{s}}(G)$. Hence (see [Ro1, §1.1]), the functor

$$\begin{aligned}\mathfrak{R}^{\mathfrak{s}}(G) &\longrightarrow \mathrm{End}_G(\Pi^{\mathfrak{s}}) \\ (\pi, \mathcal{V}) &\longmapsto \mathrm{Hom}_G(\Pi^{\mathfrak{s}}, \mathcal{V})\end{aligned} \tag{2.9.33}$$

is an equivalence of categories. This motivates the study of the endomorphism algebra $\mathrm{End}_G(\Pi^{\mathfrak{s}})$, which was carried out in [Hei, So2].

The *theory of types* is also a fruitful tool for studying the smooth complex representation theory of a reductive p-adic group G. An $\mathfrak{s}$-type (J, ϱ) in G is a pair formed by a compact open subgroup J of G and an irreducible representation ϱ of J such that containing ϱ upon restriction to J gives a necessary and sufficient condition for an irreducible representation of G to belong to $\mathrm{Irr}^{\mathfrak{s}}(G)$. Constructions of types are central to many recent developments in the representation theory of p-adic groups, in particular to the explicit constructions of supercuspidal representations, and the Bruhat–Tits building is a key ingredient of these constructions. For a comparison of the endomorphism algebra $\mathrm{End}_G(\Pi^{\mathfrak{s}})$ and the intertwining algebra $\mathcal{H}(J, \rho)$ of an $\mathfrak{s}$-type (J, ρ), see Corollary 4.5.12 in [Au3].

2.9.6 Applications of the Bruhat–Tits building in representation theory

Moy and Prasad defined in [MP1, MP2] the notion of an unrefined minimal K-type, which comes with an intrinsic depth. They proved that each smooth irreducible representation π of G contains a unrefined minimal K-type (typically not unique), and that the depth of π does not depend on the choice of unrefined minimal K-type. Thus, it defines an intrinsic invariant of the representation π. It can be defined as follows. By [DB, Lemma 5.2.1], there exists $\mathrm{d}(\pi) \in \mathbb{Q}$ with the following properties.

(1) If $(x, r) \in \mathcal{B} \times \mathbb{R}_{\geq 0}$ such that $\mathcal{V}^{G_{x,r+}}$ is nontrivial, then $r \geq \mathrm{d}(\pi)$.
(2) There exists an $x' \in \mathcal{B}$ such that $\mathcal{V}^{G_{x',\mathrm{d}(\pi)+}}$ is nontrivial.

Then the *depth* of π is defined to be the least nonnegative real number $\mathrm{d}(\pi)$ for which there exists an $x \in \mathcal{B}$ such that $V^{G_{x,\mathrm{d}(\pi)+}}$ is nontrivial. Spherical representations (or more generally Iwahori-spherical, and even more generally *unipotent* representations) have depth zero.

2.9.7 Coefficient systems on the Bruhat–Tits building

Through its partition into facets the Bruhat–Tits building $\mathcal{B}$ acquires the structure of an ℓ-dimensional locally finite polysimplicial complex (see [BT1, 1.2.1.12 and II. 5.1.32]), where $\ell := \dim V$ is the semisimple k-rank of G. For $0 \leq d \leq \ell$, we denote by $\mathcal{B}_d$ the set of all d-dimensional facets of $\mathcal{B}$. Let

$$\mathcal{B}^d := \bigcup_{F \in \mathcal{B}_d} \overline{F} \tag{2.9.34}$$

denote the d-skeleton of $\mathcal{B}$; we put $\mathcal{B}^{-1} := \emptyset$. Define an *oriented d-facet* to be, in case $d > 0$, a pair (F, c) where $F \in \mathcal{B}_d$ and c is a generator of $\mathrm{H}_d(\mathcal{B}^d, \mathcal{B}^d \backslash F; \mathbb{Z})$; then $(F, -c)$ is another oriented d-facet. An oriented 0-facet simply is a 0-facet F which we sometimes also think of as the pair $(F, 1)$, where 1 is the canonical generator of $\mathrm{H}_0(\mathcal{B}^0, \mathcal{B}^0 \backslash F; \mathbb{Z}) = \mathbb{Z}$. We denote by $\mathcal{B}_{(d)}$ the set of all oriented F-facets.

A *coefficient system* (of complex vector spaces) $\underline{V}$ on the Bruhat–Tits building $\mathcal{B}$ consists of complex vector spaces $\mathcal{V}_F$ for each facet F in $\mathcal{B}$, and linear maps

$$\mathrm{r}^F_{F'} \colon \mathcal{V}_F \longrightarrow \mathcal{V}_{F'} \quad \text{for each pair of facets } F' \subset \overline{F},$$

such that $\mathrm{r}^F_F = \mathrm{id}$ and $\mathrm{r}^F_{F'} = \mathrm{r}^{F'}_{F'} \circ \mathrm{r}^F_{F'}$ whenever $F' \subset \overline{F}'$ and $F' \subset \overline{F}$. In an obvious way the coefficient systems form a category, which we denote by $\mathrm{Coeff}(\mathcal{B})$.

When $n \in \mathbb{Z}_{\geq 0}$, the group $G_{x,n+}$ defined in (2.9.21) depends only on the facet F which contains the point x, and we will denote it by $G_{F,n+}$. In [SS], the authors consider, for any point $x \in \mathcal{B}$, the filtration

$$G_F \supset G_{F,0+} \supset G_{F,1+} \supset \cdots \tag{2.9.35}$$

of $G_F = G_{F,0}$ by the open pro-p subgroups $G_{F,n+}$ for $n \in \mathbb{Z}_{\geq 0}$.

We fix now an integer $n \geq 0$. For any representation $(\pi, \mathcal{V})$ in $\mathfrak{R}(G)$ we then have the coefficient system $\underline{V} := (\mathcal{V}^{G_{F,n+}})$ of subspaces of fixed vectors

$$\mathcal{V}_F := \mathcal{V}^{G_{F,n+}} := \left\{ v \in \mathcal{V} \,:\, \pi(g)(v) = v \quad \text{for all } g \in G_{F,n+} \right\}; \tag{2.9.36}$$

because of $G_{F',n+} \subset G_{F,n+}$ for $F' \subset \overline{F}$ the transition maps $\mathrm{r}^F_{F'}$ are the obvious inclusions.

Since the $G_{F,n}$ are profinite groups, the following functor is exact:

$$\gamma_n \colon \mathfrak{R}(G) \longrightarrow \mathrm{Coeff}(\mathcal{B}) \quad \mathcal{V} \mapsto \mathcal{V}^{G_{F,n+}}. \tag{2.9.37}$$

For any $0 \leq d \leq \ell$, the space of *oriented (cellular) d-chains* of $\gamma_n(\mathcal{V})$ is defined to be the $\mathbb{C}$-vector space $\mathrm{C}^{\mathrm{or}}_c(\mathcal{B}_{(d)}, \gamma_n(\mathcal{V}))$ of all maps of all maps $\gamma \colon \mathcal{B}_{(d)} \to \mathcal{V}$ such that γ has finite support, $\gamma((F, c)) \in \mathcal{V}^{G_{F,n+}}$, and, if $d \geq 1$, $\gamma(F, -c) = -\gamma(F, c)$, for any $(F, c) \in \mathcal{B}_{(d)}$. The group G acts smoothly on these spaces via $(g \cdot \gamma)((F, c)) := g(\gamma((g^{-1}F, g^{-1}c)))$.

The boundary map $\partial \colon \mathrm{C}^{\mathrm{or}}_c(\mathcal{B}_{(d+1)}, \gamma_n(\mathcal{V})) \longrightarrow \mathrm{C}^{\mathrm{or}}_c(\mathcal{B}_{(d)}, \gamma_n(\mathcal{V}))$ sends γ to the map

$$c(F(x'), c') \mapsto \sum_{\substack{(F,c) \in \mathcal{B}_{(d+1)} \\ F' \subset \overline{F},\, \partial^F_{F'}(c) = c'}} \gamma((F, c)). \tag{2.9.38}$$

It satisfies $\partial \circ \partial = 0$. Hence, we obtain the augmented chain complex

$$\mathrm{C}^{\mathrm{or}}_c(\mathcal{B}_{(d)}, \gamma_n(\mathcal{V})) \xrightarrow{\partial} \mathrm{C}^{\mathrm{or}}_c(\mathcal{B}_{(d-1)}, \gamma_n(\mathcal{V})) \xrightarrow{\partial} \cdots \xrightarrow{\partial} \mathrm{C}^{\mathrm{or}}_c(\mathcal{B}_{(0)}, \gamma_n(\mathcal{V})) \xrightarrow{\epsilon} \mathcal{V}, \tag{2.9.39}$$

where the augmentation map ϵ sends γ to $\sum_{F \in \mathcal{B}_{(0)}} \gamma(F)$.

For any open subgroup J of G, let $\mathfrak{R}^J(G)$ denote the category of those smooth representations $\mathcal{V}$ which are generated by their J-fixed vectors $\mathcal{V}^J$. It is a full subcategory of $\mathfrak{R}(G)$.

A point x in the Coxeter complex A is called *special* if for any direction of wall there is a wall of A actually passing through x (see [3, 1.3.7]). Let x be a special vertex in A. For every representation $\mathcal{V}$ of G in $\mathfrak{R}^{G_{F,n+}}$ the augmented complex

$$\mathrm{C}_c^{\mathrm{or}}(\mathcal{B}_{(\cdot)}, \gamma_n(\mathcal{V})) \longrightarrow \mathcal{V}$$

is an exact resolution of $\mathcal{V}$ in the category $\mathfrak{R}(G)$ (see [SS, Theorem II.3.1]). If one moreover assumes that the representation $(\pi, \mathcal{V})$ admits a central character χ, then $\mathrm{C}_c^{\mathrm{or}}(\mathcal{B}_{(\cdot)}, \gamma_n(\mathcal{V}))$ is a projective resolution of $(\pi, \mathcal{V})$ in the category of smooth representations of G with central character χ (see [SS, Corollary II.3.2], and, for a different proof in the case of Iwahori-spherical representations of the group $\mathrm{GL}_n(k)$, [Br]). These projective resolutions give rise to pseudo-coefficients for discrete series representations (generalizing the pseudo-coefficient constructed by Kottwitz in [Kot] for the Steinberg representation) as well as a Lefschetz type character formula for the Harish-Chandra character of any smooth representation.

Chapter III
Spherical functions on a group of p-adic type

Throughout this chapter and the succeeding ones, G is a simply-connected group of p-adic type, as defined in §2.7, and the notation and hypotheses of Chapter II remain in force.

3.1 The root system Σ_1

For each affine root α, the group U_α is of finite index in $U_{\alpha-1}$, by (2.7.4). We define

$$q_\alpha = (U_{\alpha-1} : U_\alpha).$$

Then it follows from axiom (I) that

$$q_{w\alpha} = q_\alpha \tag{3.1.1}$$

for all $w \in W$. In particular, since $s_{\alpha+1} \circ s_\alpha$ sends $\alpha + r$ to $\alpha + r + 2$ for all $r \in \mathbb{Z}$, we have

$$q_\alpha = q_{\alpha+2}. \tag{3.1.2}$$

Hence, if $a \in \Sigma_0$ and $r \in \mathbb{Z}$, q_{a+r} is equal to either q_a or q_{a+1}. But it can happen that $q_a \neq q_{a+1}$.

As another particular case of (3.1.1) we have

$$q_{a+r} = q_{wa+r} \quad \text{for all } w \in W_0. \tag{3.1.3}$$

We define a set Σ_1 of vectors in V^* as follows:

(1) $\Sigma_0 \subseteq \Sigma_1 \subseteq \Sigma_0 \cup \frac{1}{2}\Sigma_0$;
(2) If $a \in \Sigma_0$, then $\frac{1}{2}a \in \Sigma_1 \Leftrightarrow q_a \neq q_{a+1}$.

I. G. Macdonald and A.-M. Aubert, *Spherical Functions on a Group of p-adic Type*, Lecture Notes in Mathematics 2392,
https://doi.org/10.1007/978-3-032-15671-6_3

Proposition 3.1.4 *Σ_1 is an irreducible root system in V^*, with Weyl group W_0.*

Proof We have to show that Σ_1 satisfies the axioms of §2.1, except for (RS4). (RS1) and (RS5) follow immediately from the corresponding facts for Σ_0, and (RS2) is an immediate consequence of (3.1.3). So it remains to show that $b(c^\vee) \in \mathbb{Z}$ for all $b, c \in \Sigma_1$.

If b, c are both in Σ_0 there is nothing to prove. If $c \in \frac{1}{2}\Sigma_0$, say $c = a/2$, then $c^\vee = 2a^\vee$, and $b(2a^\vee)$ is an even integer if $b_0 \in \Sigma_0$, an integer if $b \in \frac{1}{2}\Sigma_0$. The only remaining possibility is that $c \in \Sigma_0$ and $b = a/2 \in \frac{1}{2}\Sigma_0$. Then we have to show that $a(c^\vee)$ is an even integer.

Suppose that $a(c^\vee)$ is odd. The translation t_c (defined by $t_c(0) = c^\vee$) is in the affine Weyl group, and $t_c(a) = a - a(c^\vee)$. Hence by (3.1.1) we have

$$q_a = q_{a-a(c^\vee)},$$

and therefore $q_a = q_{a+1}$ by (3.1.2), because $a(c^\vee)$ is odd. But then $b = a/2 \notin \Sigma_1$. □

For each $a \in \Sigma_0$ we define

$$q_{a/2} = q_{a+1}/q_a.$$

Thus if $b \in \Sigma_0 \cup \frac{1}{2}\Sigma_0$ we have $q_b \neq 1 \Leftrightarrow b \in \Sigma_1$. From (3.1.3), $q_{wb} = q_b$ for all $w \in W_0$.

If G_1, G_2 are subgroups of a group and $G_1 \subseteq G_2$, we write $(G_1 : G_2)$ to mean $(G_2 : G_1)^{-1}$. With this convention,

Proposition 3.1.5 $(U_{a-r} : U_a) = q_{a/2}^{[r/2]} q_a^r$, *where $[r/2]$ is the integral part of $r/2$.*

The proof is a straightforward calculation, using (3.1.2).

Let $a \in \Pi_0$. Then from (2.7.4) we have $(B : B \cap s_a B s_a) = (U_a : U_{a+1}) = q_{a+1} = q_{a/2} q_a$, so that

$$(B s_a B : B) = q_{a/2} q_a, \quad a \in \Pi_0. \tag{3.1.6}$$

For each $w \in W$, let $q(w) = (BwB : B)$.

Proposition 3.1.7 *Let $w = w_1 \cdots w_r$ be a reduced word for $w \in W$. Then $q(w) = q(w_1) \cdots q(w_r)$.*

Proof If $\alpha \in \Pi$ and $w \in W$ are such that $\ell(s_\alpha w) > \ell(w)$ then (2.3.7) $B s_\alpha w B = B s_\alpha B \cdot BwB$. Since $s_\alpha \neq w$, the cosets $B s_\alpha B$ and BwB are disjoint, and therefore $q(s_\alpha w) = q(s_\alpha) q(w)$. (3.1.7) now follows by induction on the length. □

Corollary 3.1.8 *If $w \in W_0$, then $q(w) = \prod_b q_b$, the product being taken over the $b \in \Sigma_1$ which are positive on C_0 and negative on wC_0.*

Proof Let $w = s_1 \cdots s_r$ be a reduced word for w in W_0, where $s_i = s_{a_i}$, $a_i \in \Pi_0$. Then the roots $a \in \Sigma_0$ which are positive on C_0 and negative on wC_0 are congruent under W_0 to $a_1, \ldots, a_r$. The results now follows from (3.1.7), (3.1.6) and the fact that $q_{wb} = q_b$ for all $b \in \Sigma_1, w \in W_0$. □

For each $b \in \Sigma_1$ let ξ_b be an indeterminate over $\mathbb{Z}$ such that $\xi_{wb} = \xi_b$ for all $w \in W_0$. Define

$$\xi(w) = \prod_b \xi_b \qquad (w \in W_0)$$

where as above the product is over the $b \in \Sigma_1$ which are positive on C_0 and negative on wC_0, and put

$$P_{W_0}(\xi) = \sum_{w \in W_0} \xi(w).$$

$P_{W_0}(\xi)$ is the *Poincaré polynomial* of the root system Σ_1. It is a polynomial in one variable if all the roots of Σ_1 are of the same length (types A_ℓ, D_ℓ, E_6, E_7, E_8), two variables if Σ_1 is of type B_ℓ, C_ℓ, F_4 or G_2, and three variables in the case of BC_ℓ.

Since $K = \bigcup_{w \in W_0} BwB$, we have

$$(K : B) = \sum_{w \in W_0} q(w) = P_{W_0}(q) \text{ say.} \tag{3.1.9}$$

3.2 Haar measures

If Γ is any locally compact topological group, the modulus function Δ_Γ (see e.g. [4, Chapter XIV]) is a continuous homomorphism of Γ into the multiplicative group of positive real numbers, and therefore is equal to 1 on any compact subgroup of Γ.

In the case of our group G, it follows (using (2.7.3)) from this remark that the kernel of Δ_G contains all parahoric subgroups and hence is the whole of G: in other words, G *is unimodular*. So are K and U_α, because they are compact ((2.7.3) and (2.7.4)); so also are $U_{(a)}$ and $U^\pm$, because they are unions of compact subgroups. Finally, the group Z is unimodular, for it has H as a compact normal subgroup with Z/H discrete, by (2.7.1).

On the other hand, the groups $B_0^\pm = ZU^\pm$ are not in general unimodular.

We fix Haar measures $\mathrm{d}g, \mathrm{d}k, \mathrm{d}z, \mathrm{d}u_a$ on $G, K, Z, U_{(a)}$ $(a \in \Sigma_0)$ respectively, such that

$$\int_H \mathrm{d}g = \int_K \mathrm{d}k = \int_H \mathrm{d}z = \int_{U_a} \mathrm{d}u_a = 1.$$

Next, consider the group U^+. Arrange the positive roots in a sequence $a_1, \dots, a_n$ such that height $(a_i) \le$ height (a_j) if $i < j$. Then, as we ought to have proved in §2.6, the group $U_j = \prod_{i \ge j} U_{(a_i)}$ is a normal subgroup of U^+ for $j = 1, 2, \dots, n$, and U_j is the semi-direct product of $U_{(a_j)}$ and U_{j+1}. Hence if $\mathrm{d}u_{j+1}$ is a Haar measure on U_{j+1}, it follows that $\mathrm{d}u_j = \mathrm{d}u_{a_j}\mathrm{d}u_{j+1}$ is a Haar measure on U_j. Consequently

$$\mathrm{d}u^+ = \prod_{i=1}^n \mathrm{d}u_{a_i}, \tag{3.2.1}$$

where $u^+ = \prod_{i=1}^n u_{a_i}$, is a Haar measure on $U_1 = U^+$, for which the subgroup $U_{(C_0)} = \prod_{i=1}^n U_{a_i}$ has measure 1.

The left and right Haar measures on $B_0^+ = ZU^+$ are then

$$\begin{aligned} \mathrm{d}_\ell(zu^+) &= \mathrm{d}z\, \mathrm{d}u^+ \\ \mathrm{d}_r(zu^+) &= \mathrm{d}(zu^+z^{-1})\, \mathrm{d}z = \Delta(z)\mathrm{d}z\, \mathrm{d}u^+ \end{aligned}$$

where Δ is a continuous homomorphism of Z into the multiplicative group $\mathbb{R}_+^*$ of positive real numbers, defined by $\Delta(z) = \mathrm{d}(zu^+z^{-1})/\mathrm{d}u^+$.

For any integrable function on G we have

$$\int_G f(g)\mathrm{d}g = \int_K \int_{ZU^+} f(zku^+)\mathrm{d}k\, \mathrm{d}_r(zu^+)$$

or symbolically

$$\mathrm{d}g = \Delta(z)\mathrm{d}k\, \mathrm{d}z\, \mathrm{d}u^+. \tag{3.2.2}$$

Since Z normalizes each $U_{(a)}$ we have homomorphisms $\Delta_{(a)} : Z \to \mathbb{R}_+^*$ defined by $\Delta_a(z) = \mathrm{d}(zu_a z^{-1})\mathrm{d}u_a$. Δ and Δ_a are trivial on H, hence induce homomorphisms $\delta, \delta_a : \Lambda \to \mathbb{R}_+^*$ such that $\Delta = \delta \circ \nu$, $\Delta_a = \delta_a \circ \nu$.

From (3.2.1) we have immediately

$$\Delta(z) = \prod_{a\in\Sigma_0^+} \Delta_a(z). \tag{3.2.3}$$

Also the value of $\Delta_a(z)$ is given by

Proposition 3.2.4 *Let $a \in \Sigma_0, z \in Z, t = \nu(z) \in \Lambda$. Then*

$$\Delta_a(z) = \delta_a(t) = q_{a/2}^{\frac{1}{2}a(x)} q_a^{a(x)}$$

where $x = t(0)$. If $a(x) \geq 0$ then $\delta_a(t)$ is a positive integer.

Proof Since $t(a) = a \circ t^{-1} = a - a(x)$ we have $zU_a z^{-1} = U_{a-a(x)}$ and hence

$$\begin{aligned} \Delta_a(z) &= (zU_a z^{-1} : U_a) \\ &= (U_{a-a(x)} : U_a) \\ &= q_{a/2}^{[r/2]} q_a^r \end{aligned}$$

by (3.1.5), where $r = a(x)$. Now x belongs to the lattice spanned by the $b^\vee$ $(b \in \Sigma_0)$. If $q_{a/2} \neq 1$, that is if $\frac{1}{2}a \in \Sigma_1$, then $\frac{1}{2}a(x) \in \mathbb{Z}$ by (3.1.4) and therefore r is an even integer, i.e. $[r/2] = \frac{1}{2}a(x)$. □

Corollary 3.2.5 $\Delta(z) = \delta(t) = \prod_{b\in\Sigma_1^+} q_b^{b(x)}$.

Proof (3.2.3), (3.2.4). □

Corollary 3.2.6 *If $w \in W_0$ and $n \in \nu^{-1}(w)$, then*

$$\Delta_{wa}(nzn^{-1}) = \Delta_a(z) \ (z \in Z), \quad \delta_{wa}(wtw^{-1}) = \delta_a(t) \ (t \in \Lambda).$$

Proof For $q_{wb} = q_b$ $(b \in \Sigma_1)$. □

Corollary 3.2.7 *Let $a \in \Pi_0$. Then $\delta(t_a) = \delta_a(t_a) = q_{a/2}q_a^2$.*

Proof Consider the product $\prod_b \delta_b(t_a)$ extended over the roots $b \in \Sigma_0^+, b \neq a$. Since s_a permutes this set of roots, we have

$$\prod_b \delta_b(t_a) = \prod_b \delta_{s_a b}(t_a) = \prod_b \delta_b(t_a^{-1})$$

by (3.2.6), so that $\prod_b \delta_b(t_a) = 1$. Hence $\delta(t_a) = \delta_a(t_a)$, and $\delta_a(t_a) = q_{a/2}\, q_a^2$ by (3.2.4), since $t_a(0) = a^\vee$. □

Corollary 3.2.8 *Let $t \in \Lambda^{++}, w \in W_0$. Then $\delta(t)^{1/2}/\delta(wtw^{-1})^{1/2}$ is a positive integer.*

Proof First consider the case where $w = s_a$ $(a \in \Pi_0)$, and let t be any element of Λ. Put $x = t(0)$. Then $(s_a t s_a)(0) = s_a(x) = x - a(x)a^\vee$ so that $s_a t s_a = t \cdot t_a^{-a(x)}$ and therefore

$$\begin{aligned}\delta(t)/\delta(s_a t s_a) &= \delta(t_a)^{a(x)} \\ &= q_{a/2}^{a(x)}\, q_a^{2a(x)} \quad \text{by (3.2.7)} \\ &= \delta_a(t)^2.\end{aligned} \tag{3.2.9}$$

Hence $\delta(t)^{1/2}/\delta(s_a t s_a)^{1/2}$ is a positive integer, by (3.2.4), provided that $a(x) \geq 0$.

Now let $t \in \Lambda^{++}$ and let w be any element of W_0. Let $w = s_{a_r} \cdots s_{a_1}$ be a reduced word for w, where $a_i \in \Pi_0$ $(1 \leq i \leq r)$. Put $t_i = s_{a_i} \cdots s_{a_1} t s_{a_1} \cdots s_{a_i}$ and put $x_i = t_i(0) = s_{a_i} \cdots s_{a_1}(x)$ where $x = t(0)$. Then

$$a_{i+1}(x_i) = (s_{a_1} \cdots s_{a_i} a_{i+1})(x) \geq 0$$

because $s_{a_1} \cdots s_{a_i} a_{i+1}$ is a positive root (it is one of the positive roots which is negative on $w^{-1}C_0$, see §2.1) and $x \in \overline{C_0}$. Hence from the first part of the proof we have $\delta(t_i)^{1/2}/\delta(t_{i+1})^{1/2} \in \mathbb{Z}$ and therefore $\delta(t)^{1/2}/\delta(wtw^{-1})^{1/2} = \delta(t_0)^{1/2}/\delta(t_r)^{1/2} \in \mathbb{Z}$. □

For the group U^- we have analogously

$$du^- = \prod_{i=1}^{n} du_{-a_i} \tag{3.2.1'}$$

and the right Haar measure on $B_0^- = ZU^-$ is $d_r(zu^-) = \Delta'(z)dzdu^-$, where

$$\Delta'(z) = d(zu^- z^{-1})/du^- = \prod_{a \in \Sigma_0^-} \Delta_a(z)$$

over the negative roots. By (3.2.3) and (3.2.4) this is equal to $\Delta(z)^{-1}$. Hence

$$\mathrm{d}_r(zu^-) = \Delta(z)^{-1}\mathrm{d}z\,\mathrm{d}u^- \tag{3.2.10}$$

and

$$\mathrm{d}g = \Delta(z)^{-1}\mathrm{d}k\,\mathrm{d}z\,\mathrm{d}u^-. \tag{3.2.2$'$}$$

Remark 3.2.11 In the case of a Chevalley group §2.5 all the q_a are equal to q, the number of elements in the residue field, and all the $q_{a/2}$ are equal to 1, so that $\Sigma_1 = \Sigma_0$. So in this case we have

$$\delta(t)^{1/2} = q^{r(t(0))}$$

where $r = (1/2)\sum_{a\in\Sigma_0^+} a$.

We shall need later the value of the index $v_t = (KtK : K)$, where $t \in \Lambda^{++}$. The computation fits in conveniently here.

Proposition 3.2.12 *Let C' be any chamber of Σ and let $B' = G_{(C')}$ be the corresponding Iwahori subgroup. Then $(B'tB' : B') = \delta(t)$ for all $t \in \Lambda^{++}$.*

Proof By (2.6.4) we have $B' = U^-_{(C')}HU^+_{(C')}$. If $\alpha = -a + r$ is positive on C', where $a \in \Sigma_0^+$, then $t\alpha = -a + r + a(t(0))$ is also positive on C', because $a(t(0)) \geq 0$. Hence $tU^-_{(C')}t^{-1} \subset B'$ and therefore $B'tB' = B'tU^+_{(C')}$. Consequently

$$\begin{aligned}(B'tB' : B') &= (B'tU^+_{(C')}t^{-1} : B')\\ &= (tU^+_{(C')}t^{-1} : B' \cap tU^+_{(C')}t^{-1})\\ &= (tU^+_{(C')}t^{-1} : U^+_{(C')}) \quad \text{by (2.6.5)}\\ &= \delta(t).\end{aligned}$$

□

Corollary 3.2.13 *Let $t \in \Lambda^{++}$. Then $q(wtw^{-1}) = \delta(t)$ for all $w \in W_0$.*

Proof $q(wtw^{-1}) = (Bwtw^{-1}B : B) = (B'tB' : B')$ where $B' = w^{-1}Bw$. □

Let $t \in \Lambda^{++}$ and let W_{0t} be the centralizer of t in W_0, or equivalently the subgroup of W_0 which fixes the point $t(0) \in A$. Put

$$P_{W_{0t}}(\xi) = \sum_{w\in W_{0t}} \xi(w) \tag{3.2.14}$$

in the notation of §3.1, and let $Q^t(\xi) = P_{W_0}(\xi)/W_{0t}(\xi)$ which is also a polynomial.

Proposition 3.2.15 $v_t = (KtK : K) = \delta(t)Q^t(q^{-1})$ *for all* $t \in \Lambda^{++}$.

Proof By (2.3.5) we have $KtK = BW_0tW_0B$, which is the disjoint union of the sets Bt_1W_0B as t_1 runs through the set $\{wtw^{-1} : w \in W_0\}$. For each such t_1 there exists a unique $w_1 \in W_0$ such that $t_1^{-1}C \subset w_1C_0$, i.e. $C \subset t_1w_1C_0$. Thus we have

$$(Bt_1W_0B : B) = \sum_{w\in W_0}(Bt_1w_1wB : B) = \sum_{w\in W_0} q(t_1w_1w).$$

Now by (2.2.2) and (3.1.7) we have $q(t_1w_1w) = q(t_1w_1)q(w)$. In particular, taking $w = w_1^{-1}$, it follows that $q(t_1w_1) = q(t_1)q(w_1)^{-1}$ and therefore $(Bt_1W_0B : B) = \sum_{w\in W_0} q(t_1)q(w_1)^{-1}q(w)$ or, making use of (3.2.12),

$$(Bt_1W_0B : B) = \delta(t)q(w_1)^{-1}P_{W_0}(q). \tag{3.2.16}$$

Now the subgroup W_{0t} of W_0 has a unique set W_0^t of left coset representatives with the property that

$$w^t \in W_0^t \text{ and } w_t \in W_{0t} \Rightarrow \ell(w^tw_t) = \ell(w^t) + \ell(w_t).$$

Let w_0 be the unique element of W_0 of maximum length, and let w_{0t} be the unique element of maximum length in W_{0t}. Then, with w_1 as above, we have

Lemma 3.2.17 $w_1w_0w_{0t} \in W_0^t$. □

Proof We have $t_1 = wtw^{-1}$ and we may assume that $w \in W_0^t$. Since $wt^{-1}w^{-1}C \subset w_1C_0$, we have $wt^{-1}(0) \in w_1\overline{C_0}$, hence $w_1^{-1}wt^{-1}(0) \in \overline{C_0}$; but also $w_0t^{-1}(0) \in \overline{C_0}$, because $t \in \Lambda^{++}$. Hence $w_1^{-1}w$ and w_0 both map $t^{-1}(0)$ to the same point, because $\overline{C_0}$ is a fundamental domain for W_0. It follows that $w_0w_1^{-1}w$ belongs to W_{0t}. We want to show that it is equal to w_{0t}, i.e. that it changes the sign of every $a \in \Sigma_0^+$ such that $a(t(0)) = 0$.

So let $a > 0$, $a(t(0)) = 0$ and $y \in C$. Then $z = w_1^{-1}t_1^{-1}y \in C_0$. We have

$$\begin{aligned} w_1w^{-1}wa(z) = a(t^{-1}w^{-1}y) &= a(w^{-1}y - t(0)) \\ &= wa(y) \quad \text{because } a(t(0)) = 0 \end{aligned}$$

and since $w \in W_0^t$ we have $wa > 0$, hence $wa(y) > 0$. Hence $w_1^{-1}wa > 0$ and therefore $w_0w_1^{-1}wa < 0$, as required. □

So we have $w_1w_0 = w^tw_{0t}$ for some $w^t \in W_0^t$. Since

$$\begin{aligned} \ell(w^tw_{0t}) &= \ell(w^t) + \ell(w_{0t}), \\ \ell(w_1w_0) &= \ell(w_0) - \ell(w_1) \end{aligned}$$

it follows that $q(w^t)q(w_{0t}) = q(w_0)q(w_1)^{-1}$ and therefore

$$\sum q(w_1)^{-1} = \frac{q(w_{0t})}{q(w_0)} \sum q(w^t) = Q^t(q^{-1}). \tag{3.2.18}$$

Hence finally

$$\begin{aligned}(KtK : K) &= (K : B)^{-1} \sum_{w \in W_0^t} (Bwtw^{-1}W_0B : B) \\ &= P_{W_0}(q)^{-1} \sum \delta(t)q(w_1)^{-1}P_{W_0}(q) \quad \text{by (3.2.16)} \\ &= \delta(t)Q^t(q^{-1}) \quad \text{by (3.2.18).}\end{aligned}$$

□

3.3 Zonal spherical functions on G relative to K

Let $\mathcal{H}(G, K)$ be the algebra of K-bi-invariant functions on G defined in §1.1, and let $\mathcal{H}_{\mathbb{Z}}(G, K)$ be the subring consisting of the $\mathbb{Z}$-valued functions in $\mathcal{H}(G, K)$. $\mathcal{H}_{\mathbb{Z}}(G, K)$ is called the *Hecke ring* of G relative to K. Clearly $\mathcal{H}(G, K) \cong \mathcal{H}_{\mathbb{Z}}(G, K) \otimes_{\mathbb{Z}} \mathbb{C}$.

For each $t \in \Lambda^{++}$ let χ_t be the characteristic function of the double coset KtK. By (1.1.1) and (2.6.11) the χ_t form a $\mathbb{Z}$-basis of $\mathcal{H}_{\mathbb{Z}}(G, K)$ and a $\mathbb{C}$-basis of $\mathcal{H}(G, K)$. The identity element is χ_1, the characteristic function of K.

Let $s \colon \Lambda \to \mathbb{C}^*$ be a homomorphism.[1] Then $(s\delta^{1/2}) \circ \nu$ is a continuous homomorphism of Z into $\mathbb{C}^*$, which we extend trivially across U^- to a continuous homomorphism $\phi_s \colon B_0^- \to \mathbb{C}^*$:

$$\phi_s(zu^-) = (s\delta^{1/2}, \nu(z)).$$

[To avoid large collections of brackets, we shall sometimes write (s, t) for $s(t)$, where $s \in \operatorname{Hom}(\Lambda, \mathbb{C}^*)$ and $t \in \Lambda$.] Now extend ϕ_s to a function on the whole of G by the rule

$$\phi_s(kzu^-) = (s\delta^{1/2}, \nu(z)), \;\; k \in K, \; z \in Z, \; u^- \in U^-.$$

Proposition 3.3.1 *For each* $s \in \operatorname{Hom}(\Lambda, \mathbb{C}^{\times})$ *the function*

$$\omega_s(g) = \int_K \phi_s(g^{-1}k)\mathrm{d}k$$

is a zonal spherical function on G relative to K. Moreover, if s is a character of Λ, i.e. if $|s(t)| = 1$ for all $t \in \Lambda$, then ω_s is positive definite.

Proof This is a direct consequence of the results of Chapter I. Since $K \cap B_0^- \subset HU^-$ (see the proof of (2.6.11(2)) it follows that ϕ_s restricted to B_0^- is trivial on $K \cap B_0^-$ and hence (1.2.7) is a zonal spherical function on B_0^- relative to $K \cap B_0^-$. Hence ω_s

[1] Such an s lies in $X^* \otimes_{\mathbb{Z}} \mathbb{C}$, i.e. the complex dual torus.

is a z.s.f. on G relative to K, by (1.3.2). The second assertion follows from (1.4.7) and (3.2.10). □

Proposition 3.3.2 *For all* $g_0 \in G$ *and* $s \in \mathrm{Hom}(\Lambda, \mathbb{C}^*)$, $\omega_{s^{-1}}(g_0) = \omega_s(g_0^{-1})$.

Proof Let $\overline{K} = K/(K \cap B_0^-)$ and let $\mathrm{d}\bar{k}$ be a relatively invariant measure on $\overline{K}$ with total mass 1. If $g_0^{-1}k = k'zu^-$ $(k' \in K, z \in Z, u^- \in U^-)$ then as in (1.4.6) we have $\mathrm{d}\bar{k} = \Delta(z)^{-1}\mathrm{d}\bar{k}'$, and hence

$$\begin{aligned}\omega_{s^{-1}}(g_0) &= \int_{\overline{K}} \phi_{s^{-1}}(g_0^{-1}k)\mathrm{d}\bar{k} \\ &= \int_{\overline{K}} (s^{-1}\delta^{1/2}, \nu(z))\Delta(z)^{-1}\mathrm{d}\bar{k}' \\ &= \int_{\overline{K}} (s\delta^{1/2}, \nu(z^{-1}))\mathrm{d}\bar{k}' \\ &= \int_{\overline{K}} \phi_s(g_0k')\mathrm{d}\bar{k}' = \omega_s(g_0^{-1}).\end{aligned}$$

□

The group W_0 acts on Λ by inner automorphisms, hence acts on $\mathrm{Hom}(\Lambda, \mathbb{C}^*)$: namely $(ws)(t) = s(w^{-1}tw)$, $(s \in \mathrm{Hom}(\Lambda, \mathbb{C}^*), t \in \Lambda, w \in W_0)$.

Proposition 3.3.3 *For all* $w \in W_0$ *and* $s \in \mathrm{Hom}(\Lambda, \mathbb{C}^*)$, $\omega_s = \omega_{ws}$.

This result is a consequence of the calculations we shall undertake in Chapter IV, and we shall postpone its proof to §4.5. The reader can check that there is no vicious circle involved here.

Now let $f \in \mathcal{H}(G, K)$. Then if $\hat{\omega}_s$ is the spherical measure associated with ω_s,

$$\begin{aligned}\hat{\omega}_s(f) &= (f * \omega_s)(1) \\ &= \int_G f(g)\omega_s(g^{-1})\mathrm{d}g \\ &= \int_G f(g)\left(\int_K \phi_s(gk)\mathrm{d}k\right)\mathrm{d}g \\ &= \int_G f(g)\phi_s(g)\mathrm{d}g \\ &= \int_K\int_Z\int_{U^-} f(kzu^-)(s\delta^{1/2}, \nu(z))\Delta(z)^{-1}\mathrm{d}k\,\mathrm{d}z\,\mathrm{d}u^- \\ &= \int_Z (s\delta^{1/2}, \nu(z))\left(\int_{U^-} f(zu^-)\mathrm{d}u^-\right)\mathrm{d}z.\end{aligned}$$

The integral $\Delta(z)^{-1/2}\int_{U^-} f(zu^-)\mathrm{d}u^-$ depends only on f and $\nu(z) = t$ say. We shall therefore write

$$\tilde{f}(t) = \delta(t)^{-1/2}\int_{U^-} f(zu^-)\,\mathrm{d}u^-$$

for $t \in \Lambda$ and any $z \in \nu^{-1}(t)$. Then the calculation above shows that

$$\hat{\omega}_s(f) = \sum_{t \in \Lambda} \tilde{f}(t)s(t). \tag{3.3.4}$$

The $\tilde{f}(t)$ should be regarded as the Fourier coefficients of $f \in \mathcal{H}(G, K)$. There are no convergence difficulties here: since f has compact support and the double cosets KtU^- are open (because K is open) and mutually disjoint, it follows that f takes non-zero values on only finitely many of these double cosets, i.e. $\tilde{f}$ has finite support in Λ.

Let $\mathbb{C}[\Lambda]$ be the group algebra of Λ over $\mathbb{C}$. Since Λ is a free abelian group generated by t_a with $a \in \Pi_0$, it follows that $\mathbb{C}[\Lambda] = \mathbb{C}[t_a^{\pm 1} : a \in \Pi_0]$, hence is a commutative integral domain. Consequently $\mathbb{C}[\Lambda]$ is the coordinate ring (or affine algebra) of an affine algebraic variety, say S, whose points are the $\mathbb{C}$-algebra homomorphisms $s : \mathbb{C}[\Lambda] \to \mathbb{C}$. Restricting these homomorphisms to Λ gives a bijection of S onto $\mathrm{Hom}(\Lambda, \mathbb{C}^*)$, and we shall identify $\mathrm{Hom}(\Lambda, \mathbb{C}^*)$ with S in this way.[2] Hence, by the Nullstellensatz, if $\phi \in \mathbb{C}[\Lambda]$,

$$\phi = 0 \Leftrightarrow s(\phi) = 0, \forall s \in S. \tag{3.3.5}$$

We regard the affine Weyl group W as a discrete group. Then, in the notation of §1.1, $C_c(\Lambda)$ is the algebra of functions on Λ with finite support. The mapping

$$\phi \mapsto \sum_{t \in \Lambda} \phi(t)t$$

is a $\mathbb{C}$-algebra isomorphism of $C_c(\Lambda)$ onto $\mathbb{C}[\Lambda]$. Hence $s \in S$ defines a homomorphism $C_c(\Lambda) \to \mathbb{C}[\Lambda]$, namely

$$s(\phi) = \sum_{t \in \Lambda} \phi(t)s(t).$$

The formula (3.3.4) now takes the form

$$\hat{\omega}_s(f) = s(\tilde{f}) = \sum_{t \in \Lambda} \tilde{f}(t)s(t). \tag{3.3.4'}$$

The group W_0 acts on $C_c(\Lambda)$ by transposition: $(w\phi)(t) = \phi(w^{-1}tw)$ ($\phi \in C_c(\Lambda), t \in \Lambda, w \in W_0$). Let $C_c(\Lambda)^{W_0}$ denote the subring of W_0-invariant functions in $C_c(\Lambda)$. Then we have

Theorem 3.3.6[3] *The mapping $f \mapsto \tilde{f}$ is a $\mathbb{C}$-algebra isomorphism of $\mathcal{H}(G, K)$ onto $C_c(\Lambda)^{W_0}$.*

Corollary 3.3.7 *$\mathcal{H}(G, K)$ is commutative.*

---

[2] Thus S is a complex dual torus.

[3] The mapping $\mathcal{S}: f \mapsto \tilde{f}$ is known as the *Satake transform*.

Proof First of all, let us see that $\tilde{f} \in C_c(\Lambda)^{W_0}$ whenever $f \in \mathcal{H}(G, K)$. Let $s \in S, w \in W_0$. Then

$$s(\tilde{f}) = \hat{\omega}_s(f) = \hat{\omega}_{ws}(f) = (ws)(\tilde{f}) = s(w^{-1}\tilde{f})$$

by (3.3.3) and (3.3.4′). Hence $\tilde{f} = w^{-1}\tilde{f}$ by (3.3.5).

Next, let $f_1, f_2 \in \mathcal{H}(G, K)$. Then

$$\begin{aligned} s(\widetilde{f_1 * f_2}) &= \hat{\omega}_s(f_1 * f_2) = \hat{\omega}_s(f_1)\hat{\omega}(f_2) \\ &= s(\tilde{f}_1)s(\tilde{f}_2) = s(\tilde{f}_1 * \tilde{f}_2) \end{aligned}$$

by (3.3.4′). Again by (3.3.5) we deduce that $\widetilde{f_1 * f_2} = \tilde{f}_1 * \tilde{f}_2$, and hence $f \mapsto \tilde{f}$ is a $\mathbb{C}$-algebra homomorphism.

It remains to show that the mapping $f \mapsto \tilde{f}$ of $\mathcal{H}(G, K)$ into $C_c(\Lambda)^{W_0}$ is bijective. Since the characteristic functions χ $(t \in \Lambda^{++})$ form a $\mathbb{C}$-basis of $\mathcal{H}(G, K)$, it is enough to show that the $\tilde{\chi}_t$ form a $\mathbb{C}$-basis of $C_c(\Lambda)^{W_0}$. The last two parts of (2.6.11) were designed expressly for this purpose.

For each $t \in \Lambda$, the orbit of t under W_0 (acting by inner automorphisms) has a unique representative in Λ^{++}. Let $< t >$ be the characteristic function of the W_0-orbit of t. Then these functions $< t >$, as t runs through Λ^{++}, form a $\mathbb{C}$-basis of $C_c(\Lambda)^{W_0}$. Hence we can write

$$\tilde{\chi}_t = \sum_{t'} \tilde{\chi}_t(t') < t' > \tag{3.3.8}$$

summed over $t' \in \Lambda^{++}$.

We define a total ordering on Λ as follows. Let $\Pi_0 = \{a_1, \ldots, a_\ell\}$ be the set of simple roots of Σ_0, arranged in some order. For each $t \in \Lambda$ we have $t(0) = \sum_{i=1}^{\ell} m_i a_i^\vee$ with $m_i \in \mathbb{Z}$. We define

$$\begin{cases} t \geq 0 & \Leftrightarrow \text{ the first non-zero } m_i \text{ is positive,} \\ t_1 \geq t_2 & \Leftrightarrow t_1 t_2^{-1} \geq 0. \end{cases} \tag{3.3.9}$$

In particular, the elements of Λ^{++} and $t' \in \Lambda$ we have (for any $z' \in v^{-1}(t')$)

$$\int_{Kt'U^-} \chi_t(g)\mathrm{d}g = \int_{U^-} \chi_t(z'u^-)\Delta(z')^{-1}\mathrm{d}u^- = \delta(t')^{-1/2}\tilde{\chi}_t(t').$$

Hence we have

$$\tilde{\chi}_t(t') = \delta(t')^{1/2}\mathrm{vol}(KtK \cap Kt'U^-). \tag{3.3.10}$$

So if $\tilde{\chi}_t(t') \neq 0$ it follows from (2.6.11)(3) that $tt'^{-1} \in \Lambda^+$ and hence that $t \geq t'$. Again, from (2.6.11)(4) we have $\tilde{\chi}_t(t) = \delta(t)^{1/2}$. Consequently (3.3.8) now takes the form

$$\tilde{\chi}_t = \delta(t)^{1/2} < t > + \sum_{\substack{t' < t \\ t' \in \Lambda^{++}}} \tilde{\chi}_t(t') < t' > \tag{3.3.8′}$$

from which it follows immediately (since $\delta(t)$ is never 0) that the $\tilde{\chi}_t$ form a basis of $\mathcal{H}(\Lambda)^{W_0}$. □

Remarks 3.3.11 (a) Consider the algebra $C_c(W, W_0)$. Restriction of functions to the subgroup Λ defines an isomorphism $\rho\colon C_c(W, W_0) \to C_c(\Lambda)^{W_0}$. Hence (3.3.6) provides a *canonical* isomorphism of $\mathcal{H}(G, K)$ onto $C_c(W, W_0)$, namely $f \mapsto \rho^{-1}(\tilde{f})$. (The emphasis here is on the word "canonical".)

(b) If G is a universal Chevalley group §2.5 over a non-Archimedean local field k, there is a much simpler proof that $\mathcal{H}(G, K)$ is commutative. For there exists an involutory anti-automorphism $i\colon G \to G$ such that $i(u_a(\xi)) = u_{-a}(\xi)$ for all $a \in \Sigma_0$ and $\xi \in k$, and $i(z) = z$ for all $z \in Z$. It follows that $i(K) = K$ and that i preserves Haar measure. Now let $f \in \mathcal{H}(G, K)$ and define $f^i = f \circ i$. Since i is an anti-automorphism, we have $(f_1 * f_2)^i = f_2^i * f_1^i$. On the other hand, i preserves all double cosets KzK and therefore $f = f^i$ for all $f \in \mathcal{H}(G, K)$. Hence $f_1 * f_2 = f_2 * f_1$.

The same argument will work whenever G has an anti-automorphism i such that $i(U_{a+r}) = U_{-a+r}$ for all $a \in \Sigma_0$ and $r \in \mathbb{Z}$, and $i(z) \in zH$ for all $z \in Z$. This will be the case, for example, whenever the central inversion $w_0\colon x \to -x$ is in the Weyl group W_0.[4] As a consequence this requires that $[L : \mathbb{Z}\Sigma_0] \leq 2$. This happens precisely when Σ_0 is $A_1, B_\ell, C_\ell, D_{2\ell}, E_7, E_8, F_4, G_2$. For then we may take $i(g) = n_0 g^{-1} n_0^{-1}$, where $n_0 \in \nu^{-1}(w_0)$.

(c) From (3.3.8′) it follows that the image of $\mathcal{H}_{\mathbb{Z}}(G, K)$ in $C_c(\Lambda)^{W_0}$ under $f \mapsto \tilde{f}$ is the $\mathbb{Z}$-module spanned by the elements $\delta(t)^{1/2} < t >$ with $t \in \Lambda^{++}$ (because $\delta(t')^{-1/2}\tilde{\chi}_t(t') \in \mathbb{Z}$ by (3.3.10)). If $\Sigma_0 = \Sigma_1$, then $\delta(t)^{1/2} \in \mathbb{Z}$ by (3.2.7), hence in this case the isomorphism of $\mathcal{H}(G, K)$ onto $C_c(W, W_0)$ in (a) above embeds $\mathcal{H}_{\mathbb{Z}}(G, K)$ in $\mathcal{H}_{\mathbb{Z}}(W, W_0)$, the Hecke ring of W relative to W_0.

Theorem 3.3.12

(1) *Every zonal spherical function* ω *on* G *relative to* K *is equal to* ω_s *for some* $s \in S = \mathrm{Hom}(\Lambda, \mathbb{C}^*)$.

(2) $\omega_s = \omega_{s'} \Leftrightarrow s' = ws$ *for some* $w \in W_0$.

Proof

(1) The mapping $\tilde{f} \mapsto \hat{\omega}(f)$ is a $\mathbb{C}$-algebra homomorphism $C_c(\Lambda)^{W_0} \to \mathbb{C}$, which since $C_c(\Lambda)$ is integral over $C_c(\Lambda)^{W_0}$ can be extended to a homomorphism $C_c(\Lambda) \to \mathbb{C}$. This in turn defines by restriction a homomorphism $s : \Lambda \to \mathbb{C}^*$, and we have $\hat{\omega}(f) = s(\tilde{f}) = \hat{\omega}_s(f)$ for $f \in \mathcal{H}(G, K)$. Hence $\hat{\omega} = \hat{\omega}_s$ and therefore $\omega = \omega_s$ by §1.5.

(2) Suppose that $\omega_s = \omega_{s'}$. Then s and s' agree on $C_c(\Lambda)^{W_0}$, by (3.3.4′), and therefore their kernels m, m', say, are maximal ideals of $C_c(\Lambda)$ lying over the same maximal ideal of $C_c(\Lambda)^{W_0}$. By a well-known theorem of commutative algebra (see for example [0, Chapter V, Ex. 13]) there exists $w \in W_0$ transforming m into m'. Hence $s' = ws$. □

[4] It is known that $-1 \in W_0$ if and only if $\mathbb{Z}\Sigma_0 \supset 2L$, where Σ_0 is the irreducible root lattice, $\mathbb{Z}\Sigma_0$ is the root lattice, and L is the weight lattice.

Remark 3.3.13 Since $\mathbb{C}^* = \mathbb{R} \times (\mathbb{R}/\mathbb{Z})$ (the isomorphism being $re^{2\pi i\theta} \mapsto (\log r, \theta)$) we have $S = \text{Hom}(\Lambda, \mathbb{C}^*) \cong \text{Hom}(\Lambda, \mathbb{R}) \times \text{Hom}(\Lambda, \mathbb{R}/\mathbb{Z})$. Now $\text{Hom}(\Lambda, \mathbb{R})$ can be identified with V^*: if $\lambda \in \text{Hom}(\Lambda, \mathbb{R})$, the corresponding element of V^* is the linear form which maps $t(0)$ to $\lambda(t)$ for all $t \in \Lambda$. This identification induces an identification of $\text{Hom}(\Lambda, \mathbb{R}/\mathbb{Z})$ with V^*/L, where (as in §2.5) L is the lattice of linear forms u on A such that $u(a^\vee) \in \mathbb{Z}$ for all $a \in \Sigma_0$. Hence the space Ω of z.s.f. may be identified with the orbit space $(V^* \times (V^*)/L))\,/W_0$. Alternatively, Ω may be identified with S/W_0, the affine algebraic variety whose coordinate ring is $C_c(\Lambda)^{W_0} \cong \mathbb{C}[\Lambda]^{W_0}$.

Theorem 3.3.14 *Let $t_1, \ldots, t_m$ be a set of generators of the semigroup Λ^{++}. Then $\mathcal{H}_\mathbb{Z}(G, K) = \mathbb{Z}[\chi_{t_1}, \ldots, \chi_{t_m}]$. (Of course, in general the χ_{t_i} will not be algebraically independent.)*

Proof For each $t \in \Lambda^{++}$ let $\theta_t = \delta(t)^{1/2} < t >$ so that the image of $\mathcal{H}_\mathbb{Z}(G, K)$ under $f \mapsto \tilde{f}$ is the $\mathbb{Z}$-module freely generated by the θ_t, by (3.3.11) (c). Let $M_t = \sum_{t'<t} \mathbb{Z}\theta_{t'}$. From (3.3.8′) we have $\tilde{\chi}_t \equiv \theta_t (\bmod M_t)$.

Also $\theta_{t_1} * \theta_{t_2} \equiv \theta_{t_1 t_2} (\bmod M_{t_1 t_2})$. For $\theta_{t_1} * \theta_{t_2} - \theta_{t_1 t_2}$ is a sum of terms of the form

$$\delta(t_1 t_2)^{1/2} \delta(t_3)^{-1/2} \theta_{t_3}$$

with $t_3 = t_1' t_2' \in \Lambda^{++}$ and $t_1' = w_1 t_1 w_1^{-1}, t_2' = w_2 t_2 w_2^{-1}$ $(w_1, w_2 \in W_0)$ and either $t_1' \neq t_1$ or $t_2' \neq t_2$. By (3.2.8) the coefficient $\delta(t_1 t_2)^{1/2} \delta(t_3)^{-1/2}$ is a rational integer, and we have $t_3 < t_1 t_2$.

Hence if $t = \prod_{i=1}^m t_i^{\mu_i} \in \Lambda^{++}$ it follows that

$$\tilde{\chi}_t \equiv \theta_t \equiv \prod_{i=1}^m \theta_{t_i}^{\mu_i} \equiv \prod_{i=1}^m \tilde{\chi}_{t_i}^{\mu_i} (\bmod M_t)$$

and so, by induction on t, that $\tilde{\chi}_t \in \mathbb{Z}[\tilde{\chi}_{t_1}, \ldots, \tilde{\chi}_{t_m}]$. Hence $\chi_t \in \mathbb{Z}[\chi_{t_1}, \ldots, \chi_{t_m}]$. □

3.4 Notes on Chapter III

This chapter is for the most part a transcription into the present context of Satake's paper [14]. In particular, theorems (3.3.6), (3.3.12) and (3.3.14) are due to him.

3.5 Additions to Chapter III

Note We put $S_+ = \text{Hom}(\Lambda, \mathbb{R})$ (called the *real split* part of S), and put $S_u = \text{Hom}(\Lambda, \mathbb{R}/\mathbb{Z})$ (called the *unitary* (or *compact*) part of S). We have in general $V = \mathbb{R} \otimes_\mathbb{Z} X_*$ and $V^* = \mathbb{R} \otimes_\mathbb{Z} X^*$, where X_* (resp. X^*) is the cocharacter (resp. character) lattice of the maximal torus of G. In the present case X_* is the coroot lattice Λ and

X^* is the weight lattice L, because $\mathbf{G}$ is simply connected. We can interpret the isomorphism $S_u \cong \mathrm{Hom}(\Lambda, \mathbb{R}/\mathbb{Z})$ using the following exact sequence

$$1 \to L \to V^* = \mathbb{R} \otimes_{\mathbb{Z}} X^* \xrightarrow{\exp} S_u \to 1,$$

where the exponential map is defined by $a^\vee(\exp(z)) = e^{2\pi i \langle a^\vee, z\rangle}$ for every $a \in \Sigma_0, z \in V^*$. Here $a^\vee$ on the left-hand side is viewed as a character of S restricted to S_u, and on the right-hand side $a^\vee$ is a linear functional on V^*.

3.5.1 Poincaré polynomial and Poincaré series

Let (W, S) be a Coxeter system with S finite. Denote by S_i $(i \in I)$ the equivalence classes for the relation "s is conjugate to s' in W" between elements s, s' of S. Let $\xi = (\xi_i)_{i\in I}$ be a family of indeterminates indexed by I, and for each $s \in S$ define ξ_s to be ξ_i if $s \in S_i$. Let w be an element of W and let $w = s_1 \cdots s_r$ be a reduced decomposition of w. The monomial $\xi_w := \xi_{s_1} \cdots \xi_{s_r}$ does not depend on the choice of the reduced decomposition (see [1, Proposition 5]).

The *Poincaré series* of (W, S) is defined to be

$$P_W(\xi) := \sum_{w\in W} \xi_w. \tag{3.5.1}$$

It is a formal power series (or polynomial, if W is finite) in the ξ_i with positive integer coefficients, which represents a rational function of ξ_i (see [1, Chapter IV, Exercise 26)]). It is proved in [Mac1] that if W is either finite or affine, the rational function $P_W(\xi)$ can be expressed quite explicitly as a product of factor of the form $(1 - m_\alpha(\xi))^{\pm 1}$, where the m_α are monomials in the ξ_i.

3.5.2 The Satake transform

Let now $\mathbf{G}$ be an arbitrary connected reductive group over k. Let x be a special vertex in the Bruhat–Tits building $\mathcal{B}(\mathbf{G}_{\mathrm{ad}}(k))$, and let K_x denote a special maximal compact open subgroup of $G := \mathbf{G}(k)$ which fixes x. Whenever $\mathbf{G}$ is *unramified* (i.e., G is quasi-split and split over an unramified extension of k), let $\mathbf{T}$ denote a maximal k-split torus in $\mathbf{G}$ whose corresponding apartment in $\mathcal{B}(\mathbf{G}_{\mathrm{ad}}(k))$ contains x, then the Satake transform is a $\mathbb{C}$-algebra isomorphism

$$\mathcal{H}(G, K) \xrightarrow{\cong} \mathbb{C}[X_*(T)]^{W_0}, \tag{3.5.2}$$

where $W_0 = W(G, A)$ is the relative Weyl group (see [Car]).

3.5.3 Spherical representations of p-adic groups

Recall that K is a maximal compact subgroup of p-adic group G. If $(\pi, \mathcal{V})$ is a smooth irreducible representation of G, then the subspace $\mathcal{V}^K$ of K-fixed vectors is at most one-dimensional (see [Car, p. 152]). The smooth irreducible representations of G which contain a non-zero K-fixed vector are said to be *K-spherical* (or *K-unramified*). Hence for a K-spherical representation $(\pi, \mathcal{V})$ of G, the vector space $\mathcal{V}^K$ is one-dimensional: $\mathcal{V}^K = \mathbb{C} \cdot v_0$ for some non-zero $v_0 \in \mathcal{V}$. The vector v_0 is called a *spherical vector*.

The irreducible unramified representations of G are in bijection with the spectrum of the spherical Hecke algebra $\mathcal{H}(G, K)$.

Remark The notion of spherical representation is motivated by the study of automorphic representations. Let F be a global field, that is, F is either a finite extension of $\mathbb{Q}$ (a number field) or a finite separable extension of $\mathbb{F}_q(t)$ (a function field). Let $\mathbb{A}_F$ denote the adele ring of F. For any finite set S of places of F, let $\mathbb{A}_F^S$ be the restricted direct product of F_v's over all $v \notin S$, where F_v denotes the completion of F at v. Any irreducible admissible representation π of $\mathbf{G}(\mathbb{A}_F)$ is uniquely a restricted tensor product of irreducible smooth representations π_v of $\mathbf{G}(F_v)$, with π_v spherical with respect to a *hyperspecial* (see for instance [Ti2]) maximal compact open subgroup K_v, for almost all finite places v (see [Fla]).

Let $(\pi, \mathcal{V})$ be a representation of G which is unitary with respect to an inner product $\langle\ ,\ \rangle$. Given vectors v and v' in $\mathcal{V}$, the corresponding *matrix coefficient* of π is defined to be the function $\pi_{v,v'}$ on G defined by

$$\pi_{v,v'} : g \mapsto \langle \pi(g)v, v' \rangle, \quad \text{for } g \in G. \tag{3.5.3}$$

If $(\pi, \mathcal{V})$ is K-spherical, then its contragredient $(\tilde{\pi}, \widetilde{\mathcal{V}})$ is also K-spherical. If $v \in \mathcal{V}$ and $\tilde{v} \in \widetilde{\mathcal{V}}$ satisfy $\langle v, \tilde{v} \rangle = 1$, then the function $\pi_{v,\tilde{v}}$ is a spherical function.

An irreducible unitary representation with square integrable matrix coefficients is called a *square integrable representation*. The *discrete series* of representations is the collection of square integrable (mod the center of G) representations. It splits into two classes: supercuspidal representations: irreducible unitary representations whose matrix coefficients are compactly supported (mod the center of G), and generalized special representations: irreducible unitary representations whose matrix coefficients are square-integrable (mod the center of G), and which are subrepresentations of representations induced from proper parabolic subgroups of G.

Suppose that G is split adjoint, and fix a Borel subgroup B of G with unipotent radical U. We consider its Levi decomposition $B = TU$, where T is the maximal split torus. Furthermore, we fix a maximal compact subgroup K of G. Let ψ be a smooth complex character of U. It factors through

$$U/[U, U] = \prod_{a \in -\Pi_0} U_a/U_{2a}, \tag{3.5.4}$$

where $U_{2a} := \{1\}$ if $2a \notin \Sigma_0$. Thus to define ψ it suffices to define smooth characters $\psi_a : U_a \to \mathbb{C}^\times$ for all $a \in -\Pi_0$, and set $\psi := \prod_{a \in -\Pi_0} \psi_a$.

Definition 3.5.5

1. The character ψ is *non-degenerate* if $\psi|_{U_a}$ is not trivial for all $a \in -\Pi_0$.
2. A *Whittaker function* is a function $\mathcal{W} : G \to \mathbb{C}$ such that for all $u \in U$ and $g \in G$, we have $\mathcal{W}(ug) = \psi(u)\mathcal{W}(g)$ with ψ a non-degenerate character of U.

Suppose that ψ is non-degenerate. For an irreducible representation π of G it is well known that

$$\dim \operatorname{Hom}(\pi, \operatorname{Ind}_U^G \psi) \leq 1, \tag{3.5.6}$$

and if (3.5.6) is non-zero we say that the representation π is *generic*. The *Whittaker model* of such a generic irreducible representation π of G is the image of an embedding $\pi \hookrightarrow \operatorname{Ind}_U^G \psi$.

Let now π be a generic irreducible representation of G. The explicit formula for the function $\mathrm{W}(v_0)$, with v_0 a spherical vector, was given in [CS] and is known as the *Casselman–Shalika formula*. The Casselman–Shalika formula [CS, Theorem 5.4] is a beautiful formula relating values of special functions on G to the values of finite-dimensional complex representations of its dual group. Further, this formula is particularly useful in the theory of automorphic forms for studying L-functions. It allows the computation of explicit formulas for periods of irreducible unramified representations of G that are associated to unique models (i.e., multiplicity-free induced representations), see notably [Sa1].

Casselman and Shalika's original approach in [CS] is based on decomposing the spherical Whittaker function using Casselman's basis. One function in this decomposition can be computed directly; the others can then be computed via a symmetry. One can manipulate the resulting formula and apply the Weyl character formula to find a character of a finite-dimensional representation of the Langlands dual group of G.

A proof of the Casselman–Shalika formula that does not use the uniqueness of the Whittaker model was obtained by N. Gurevich in [Gur], in the case where $\mathbf{G}$ is split of adjoint type, via a description of the Whittaker spherical space $(\operatorname{ind}_U^G \psi)^K$ as a concrete $\mathcal{H}(G, K) \simeq \mathbb{C}[\Lambda]^{W_0}$-module, showing that there is a canonical isomorphism from $(\operatorname{ind}_U^G \psi)^K$ to the space of functions on the lattice of coweights, that are skew-invariant under the action of W_0, which is compatible with the Satake isomorphism. The element $f \in \mathbb{C}[\Lambda]$ is called skew-invariant if $w(f) = (1)^{\ell(w)}$, where $\ell(w)$ is the length of $w \in W_0$. It follows that the *twisted Satake transform* $\mathcal{S}_\psi : \mathcal{C}(G/K) \to (\operatorname{ind}_U^G \psi)^K$, defined by

$$\mathcal{S}_\psi(f)(t) := \int_U f(ut)\overline{\psi(u)}\mathrm{d}u, \tag{3.5.7}$$

sends the spectral basis of $\mathcal{H}(G, K)$ to the basis of characteristic functions of $\operatorname{ind}_U^G \psi)$. By [FGKV], this result is equivalent to the Casselman–Shalika formula in [CS]. Later, in [GK], Gurevich and Karasiewicz established the Casselman–Shalika formula for arbitrary unramified groups over a non-Archimedean local field by studying the action of the spherical Hecke algebra on the space of compact spherical Whittaker functions via the twisted Satake transform. This method provides a conceptual explanation of the appearance of characters of a dual group in the Casselman–Shalika formula.

3.5.4 The modulo p Satake transform

Let K be an open compact subgroup of a p-adic group G. We suppose that K is a special parahoric subgroup of G. The quotient $\mathbb{G}$ of K by its pro-p unipotent radical is the group of $\mathbb{F}_q$-points of a connected reductive group defined over $\mathbb{F}_q$. Let C be some commutative field of characteristic p (for instance we can take for C an extension of $\mathbb{F}_q$ containing a root of unity of order the prime to p part of the exponent of $\mathbb{G}$). Let π be a smooth representation of G on a C-vector space V. We view V as a representation of K, and we define $\mathcal{H}_C(G, K, V)$ to be the algebra of intertwining operators of the compactly induced representation $\mathrm{c-Ind}_K^G V$. It is the algebra of compactly supported functions $f\colon G \longrightarrow \mathrm{End}_C(V)$ satisfying

$$f(k_1 g k_2) = k_1 f(g) k_2 \quad \text{for all } k_1, k_2 \in K \text{ and } g \in G,$$

and the multiplication is given by the convolution

$$(f_1 * f_2)(g) := \sum_{x \in G/K} f_1(x) \circ f_2(x^{-1}g) \quad \text{for } g \in G.$$

The algebra $\mathcal{H}_C(G, K, V)$ need not be commutative (see [HV, Proposition 2.4]); this is in contrast with the classical situation where V is trivial and $C = \mathbb{C}$.

The general framework for Satake transforms is the following. We fix a minimal parabolic subgroup P_0 of G and a Levi decomposition $P_0 = L_0 U_0$, where U_0 is the unipotent radical of P_0. We assume that $G = P_0 K$ and that $P_0 \cap K = (L_0 \cap K) \cdot (U_0 \cap K)$. Henniart and Vignéras constructed in [HV, Definition 5.3 and Proposition 5.4] a natural algebra homomorphism, the *mod p Satake transform*:

$$\mathcal{S}\colon \mathcal{H}_C(G, K) \longrightarrow \mathcal{H}_C(L_0, L_0 \cap K, V^{U_0 \cap K}). \tag{3.5.8}$$

It is given by

$$\mathcal{S}(f)(\ell) = \sum_{u \in U_0/U_0 \cap K} f(\ell u)|_{V^{U_0 \cap K}}, \quad \text{for all } \ell \in L_0. \tag{3.5.9}$$

The natural image in $\mathbb{G}$ of $P_0 \cap K$ is the group of points of a minimal parabolic subgroup $\mathbb{P}_0 = \mathbb{L}_0 \mathbb{U}_0$ of $\mathbb{G}$. The space $V^{\mathbb{U}_0}$ is one-dimensional with a natural action of $\mathbb{L}_0$ by a character $\chi_V\colon \mathbb{L}_0 \to C^\times$. We denote by $L_{0,V}^-$ the monoid of antidominant elements of L_0 that normalize the character χ_V. The image of S vanishes outside

$L^-_{0,V}$ (see [HV, Proposition 5.5]). Then $\mathcal{S}$ induces an isomorphism

$$\mathcal{H}_C(G, K, V) \xrightarrow{\sim} \mathcal{H}_C(L^-_{0,V}, L_0 \cap K, V^{U_0 \cap K}), \tag{3.5.10}$$

where $\mathcal{H}_C(L^-_{0,V}, L_0 \cap K, V^{U_0 \cap K})$ is the subalgebra of $\mathcal{H}_C(L_0, L_0 \cap K, V^{U_0 \cap K})$ of functions with support contained in $L^-_{0,V}$ (see [HV, Theorem 5.6]).

3.5.5 The geometric Satake equivalence

Let $\mathbf{G}$ be a reductive algebraic group, and let $\mathbf{G}^\vee$ be the reductive algebraic group with root datum dual to that of $\mathbf{G}$ (which means that it is obtained from that of $\mathbf{G}$ by exchanging weights and coweights and roots and coroots); this group is defined up to isomorphism. We set $G := \mathbf{G}(\mathbb{C})$ and $G^\vee := \mathbf{G}^\vee(\mathbb{C})$. Let $\mathfrak{R}_\mathbb{C}(G^\vee)$ be the category of finite-dimensional complex representations of $G^\vee$. The character map defines an isomorphism of algebras $\mathfrak{R}_\mathbb{C}(G^\vee) \cong \mathbb{C}[\Lambda]^{W_0}$.

The work of Kazhdan and Lusztig [KL1, KL2] shows that the representation theory of $G^\vee$ is closely related to the singularities arising in certain orbit closures inside affine flag varieties associated to G. Building upon [Lus1], Mirković and Vilonen established in [MV] an equivalence of symmetric tensor categories between the category $\mathfrak{R}_\mathbb{C}(G^\vee)$ and the category of certain sheaves on an infinite-dimensional variety Gr_G known as the *affine Grassmannian* of G.

We will recall the definition of $\mathrm{Gr}_{G^\vee}$, following [BR]. Let $\mathfrak{O} := \mathbb{C}[[t]]$ and $K := \mathbb{C}((t))$, where t is an indeterminate. If H is a linear complex algebraic group, we denote by $H_\mathfrak{O}$ and H_K the functors from $\mathbb{C}$-algebras to groups defined by

$$H_\mathfrak{O} \colon H \mapsto H(R[[t]]) \quad \text{and} \quad H_R \colon H \mapsto H(R((t))). \tag{3.5.11}$$

Then $H_\mathfrak{O}$ is represented by a $\mathbb{C}$-group scheme (not of finite type in most cases), and H_K by an ind-group scheme (i.e., an inductive limit of schemes parametrized by $\mathbb{Z}_{\geq 0}$, with closed embeddings as transition maps). We still denote these (ind-)group schemes by $H_\mathfrak{O}$ and H_K.

We fix a pair $\mathbf{B} \supset \mathbf{T}$ of a Borel subgroup of $\mathbf{G}$, and a maximal torus $\mathbf{T}$. Let $L^{\leq 0}G$ denote the ind-group scheme which represents the functor $R \mapsto R(G[t^{-1}])$, and let $L^{>0}G$ be the kernel of the natural morphism $L^{\leq 0}G \to G$, where $t^{-1} \mapsto 0$. Then $L^{>0}G$ is a subgroup of G_K in a natural way, and the multiplication morphism

$$L^{>0}G \times G_\mathfrak{O} \to G_K$$

is an open embedding. The quotient

$$\widetilde{\mathrm{Gr}}_G := G_K / G_\mathfrak{O} \tag{3.5.12}$$

has a natural structure of an ind-scheme. This ind-scheme is ind-proper and of ind-finite type, but not necessarily reduced. We define Gr_G to be the (reduced) ind-variety associated with the ind-scheme $\widetilde{\mathrm{Gr}}_G$.

Let $\mathrm{Perv}_{G_{\mathfrak{O}}}(\mathrm{Gr}_{G^\vee})$ denote the category of $G_{\mathfrak{O}}$-equivariant perverse sheaves on Gr_G. The *geometric Satake equivalence* is the statement that the category $\mathrm{Perv}_{G_{\mathfrak{O}}}(\mathrm{Gr}_G)$ is equivalent to the category $\mathfrak{R}_{\mathbb{C}}(G^\vee)$, in such a way that the tensor product of representations corresponds to a natural operation in $\mathrm{Perv}_{G_{\mathfrak{O}}}(\mathrm{Gr}_G)$ called convolution. This categorical equivalence is a vast generalization of the classical Satake isomorphism (via the sheaf-to-function dictionary). It is a very important tool in geometric representation theory which appears in different contexts, has a wide range of applications, and is one of the cornerstones of the geometric Langlands program. This equivalence holds even for more general coefficients such as, for instance, $\mathbb{Z}$ or $\mathbb{F}_p$. For more results on this equivalence, one may notably consult [BR], [Zhu], [MV] and [Ri].

3.5.6 Kac–Moody groups, hovels, and spherical functions

Kac–Moody groups are interesting infinite-dimensional generalizations of reductive groups, that generalize reductive groups. In order to study them over fields endowed with a discrete valuation, Gaussent and Rousseau introduced *masures* (also known as hovels) in [GaRo1], which are analogues of Bruhat–Tits buildings.

A *Kac–Moody matrix* (or *generalized Cartan matrix*) is a square matrix $C = (c_{i,j})_{i,j\in I}$, with integer coefficients, indexed by a finite set I, satisfying:

1. $c_{i,i} = 2$ for all $i \in I$;
2. $c_{i,j} \leq 0$, for all $i, j \in I$ such that $i \neq j$;
3. $c_{i,j} = 0$ if and only if $c_{j,i} = 0$.

A *root generating system* is a 5-tuple $S = (C, X, Y, (a_i)_{i\in I}, (a_i^\vee)_{i\in I})$ formed by a Kac–Moody matrix C indexed by I, of two dual free $\mathbb{Z}$-modules X (of characters) and Y (of cocharacters) of finite rank, a family $(a_i)_{i\in I}$ (of simple roots) in X and a family $(a_i^\vee)_{i\in I}$ (of simple coroots) in Y, which satisfy $c_{i,j} = a_j(a_i^\vee)$ for all $i, j \in I$. The *Langlands dual* of S is defined to be $S^\vee := ({}^tC, Y, X, (a_i^\vee)_{i\in I}, (a_i)_{i\in I})$, where ${}^tC$ is the transposed matrix of C.

Let $V := Y \otimes_{\mathbb{Z}} \mathbb{R}$. Every element in X defines a linear form on V. For $i \in I$, the formula $s_i(v) := v - a_i(v)a_i^\vee$ defines an involution in V, which is a reflection through the hyperplane $\ker(a_i)$. Let W_0 be the subgroup of $\mathrm{GL}(V)$ generated by the set $\{s_i\}_{i\in I}$. Then W_0 is a Coxeter group; it stabilizes the lattice Y of V. The positive fundamental chamber

$$\mathcal{C}_0 := \{v \in V \; :, a_i(v) > 0 \;\; \text{for all } i \in I\} \tag{3.5.13}$$

is a nonempty open convex cone. Its closure $\overline{\mathcal{C}}_0$ is the disjoint union of the *vectorial faces* $F_0(J)$ for $J \subset I$, where

$$F_0(J) := \{v \in V \,:\, a_i(v) = 0 \text{ for all } i \in J, \text{ and } a_i(v) > 0 \text{ for all } i \in I \backslash J\}\,. \tag{3.5.14}$$

The *Tits cone* $\mathcal{T}$ is the (disjoint) union of the positive vectorial faces. It is a W_0-stable convex cone in V. For $x \in \mathcal{T}$, we denote by x^{++} the unique element in $\overline{\mathcal{C}}_0$ which is conjugated by W_0 to x.

The *Kac–Moody algebra* $\mathfrak{g}$ is a complex Lie algebra generated by the standard Cartan subalgebra $\mathfrak{h} := Y \otimes \mathbb{C}$ and the Chevalley generators $(e_i)_{i \in I}$, $(f_i)_{i \in I}$. Let Σ denote the set of roots of $\mathfrak{g}$ (with respect to $\mathfrak{h}$). A root $a \in \Sigma$ is called *real* if $a = w(a_i)$ for some $w \in W_0$. Let Σ_{re} be the set of real roots. The set Σ_{re} is infinite except in the classical case, where C is a Cartan matrix and $\mathfrak{g}$ is finite-dimensional (reductive). We assume that the Lie algebra $\mathfrak{g}$ is *symmetrizable*, that is, endowed with a nondegenerate invariant $\mathbb{C}$-valued symmetric bilinear form.

Let $\mathbf{G}$ be the (split, complex) Kac–Moody group associated to the above root-generating system as defined by Tits in [Ti4] (see also [Mar, Ré]). For any field F containing $\mathbb{C}$, the group $G := \mathbf{G}(F)$ of the F-rational points of $\mathbf{G}$ is generated by the fundamental torus $\mathbf{T}(F)$, where $\mathbf{T} = \mathrm{Spec}(\mathbb{Z}[X])$, and the root subgroups $\mathbf{U}_a(F)$ for $a \in \Sigma_{\mathrm{re}}$, each isomorphic to the additive group $(F, +)$ by an isomorphism (of algebraic groups). In [BrKa], Braverman and Kazhdan defined spherical Hecke algebras for unramified split affine Kac–Moody groups over a local non-Archimedean field and proved a generalization of the Satake isomorphism for these algebras.

We assume that K is a discretely valuated field with residue field containing $\mathbb{C}$. Then, in [GaRo1], Gaussent and Rousseau associated to G a new remarkable geometric object $\mathcal{I}(G)$, called a *hovel* (*masure* in French). In the special case when $\mathbf{G}(\mathbb{C})$ is a finite-dimensional reductive group and $F = \mathbb{C}((t))$, the hovel coincides with the Bruhat–Tits building associated with G. In general, the hovel is likewise a space covered by subsets called *apartments* and endowed with an action of G permuting the apartments transitively, but it is no longer true that any two points are contained in a common apartment, a weakening of the axioms that reflects the failure of a Cartan decomposition for G. The apartments are in one-to-one correspondence with the maximal split subtori, hence permuted transitively by G. Each apartment $\mathcal{A}$ is a finite-dimensional real affine space, and its stabilizer N in G acts on it via an extended affine Weyl group $W^{\mathrm{e}} = W_0 \rtimes Y$, which stabilizes a set of affine hyperplanes called *walls*. Since W_0 is generally infinite, the extended affine Weyl group does not act properly discontinuously on V. This is reflected by the fact that $\mathcal{I}$ does not carry any simplicial structure in general. The hovel $\mathcal{I}$ is endowed with a G-invariant preorder $\leq$ that induces the known one on $\mathcal{A}$. If $x \leq y$, then x and y are in a same apartment $g \cdot \mathcal{A}$. We set $\mathcal{I}_0 := G/K$ and define

$$\mathcal{I}_0 \times_{\leq} \mathcal{I}_0 := \{(x, y) \in \mathcal{I}_0 \times \mathcal{I}_0 \,:\, x \leq y\}\,. \tag{3.5.15}$$

The vectorial distance $d_0(x, y) \in \overline{\mathcal{C}}_0$ is defined by $d_0(x, y) := (g^{-1}y - g^{-1}x)^{++}$. It does not depend on the choices we made. The elements of Y, through the identification $Y = N \cdot 0$, are called *vertices of type* 0 in $\mathcal{A}$; they are special vertices.

Let K be the stabilizer of a special point 0 in a chosen standard apartment of $\mathcal{I}$. We define the semigroup G^+ by

$$G^+ := \{g \in G \, : \, 0 \le g \cdot 0\} \,. \tag{3.5.16}$$

We set $Y^{++} := \{y^{++} \, : \, y \in Y\}$. Via d_0, the space of G-invariant functions on $\mathcal{I}_0 \times_{\le} \mathcal{I}_0$ with values in $\mathbb{C}$ is linearly isomorphic to the space

$$\widehat{\mathcal{H}}(G, K) := \left\{f : Y^{++} \simeq K \backslash G^+ / K \to \mathbb{C}\right\}, \tag{3.5.17}$$

which can be interpreted as the space of K-bi-invariant functions on G^+.

Let $\mathcal{H}(G, K)$ be the subspace of $\widehat{\mathcal{H}}(G, K)$ of functions on G with *almost finite support*, that is, satisfying

$$\operatorname{supp}(f) \subset \bigcup_{i \in I} (\lambda_i - Q_+^\vee) \cap Y^{++}, \tag{3.5.18}$$

where $\lambda_i \in Y^{++}$, and $Q_+^\vee$ is the subsemigroup of Y generated by the fundamental coroots. In [GaRo2], Gaussent and Rousseau proved that the structure constants of $\mathcal{H}(G, K)$ are polynomials depending on the geometry of the standard apartment, and established a version of the Satake isomorphism between $\mathcal{H}(G, K)$ and the algebra of Weyl invariant elements in a certain completion of an algebra of Laurent polynomials.

Independently, Braverman and Kazhdan introduced in [BrKa] the spherical Hecke algebra and the Satake isomorphism for an untwisted affine Kac–Moody group over a non-Archimedean local field. Jointly with Patnaik, they further developed in [BKP] the theory of the Iwahori–Hecke algebras associated to these groups and constructed the affine Satake isomorphism, which generalizes Macdonald's formula for spherical functions in the finite-dimensional case.

Chapter IV
Calculation of the spherical functions

4.1 Statement of result

For each $b \in \Sigma_0 \cup \frac{1}{2}\Sigma_0$ let $t_b \in \Lambda$ be the translation defined by $t_b(0) = b^\vee$ (so that $t_{a/2} = t_a^2 = t_a \circ t_a$). Let $s \in \mathrm{Hom}(\Lambda, \mathbb{C}^*)$ and define

$$c(b,s) = \frac{1 - q_{b/2}^{-1/2} q_b^{-1} s(t_b)^{-1}}{1 - q_{b/2}^{-1/2} s(t_b)^{-1}}$$

whenever the denominator is non-zero. (If $b \notin \Sigma_1$, then q_b is taken to be 1. Likewise for $q_{b/2} = q_{b+1}/q_b$.) If $a \in \Sigma_0$ we put

$$c_0(a,s) = c(\frac{1}{2}a, s)c(a,s).$$

So if $\frac{1}{2}a \notin \Sigma_1$ we have

$$c_0(a,s) = c(a,s) = \frac{1 - q_a^{-1} s(t_a)^{-1}}{1 - s(t_a)^{-1}}$$

whilst if $\frac{1}{2}a \in \Sigma_1$ we have

$$\begin{aligned} c_0(a,s) &= \frac{1 - q_{a/2}^{-1} s(t_a)^{-2}}{1 - s(t_a)^{-2}} \cdot \frac{1 - q_{a/2}^{-1/2} q_a^{-1} s(t_a)^{-1}}{1 - q_{a/2}^{-1/2} s(t_a)^{-1}} \\ &= \frac{(1 + q_{a/2}^{-1/2} s(t_a)^{-1})(1 - q_{a/2}^{-1/2} q_a^{-1} s(t_a)^{-1})}{1 - s(t_a)^{-2}}. \end{aligned}$$

Finally put

$$c(s) = \prod_{b \in \Sigma_1^+} c(b,s) = \prod_{a \in \Sigma_0^+} c_0(a,s).$$

I. G. Macdonald and A.-M. Aubert, *Spherical Functions on a Group of p-adic Type*,
Lecture Notes in Mathematics 2392,
https://doi.org/10.1007/978-3-032-15671-6_4

Then $c(s)$ is defined whenever $s(t_b) \neq 1$ for all $b \in \Sigma_1$. The function c is the counterpart of Harish-Chandra's c-function for a real semisimple Lie group.

Lemma 4.1.1 (1) $c(b,s) = c(wb, ws)$ *for* $b \in \Sigma_1$ *and* $w \in W_0$.
(2) $c_0(a,s) + c_0(-a,s) = 1 + q(s_a)^{-1}$.
(3) $c(s^{-1}) = c(w_0 s)$, *where* w_0 *is the element of* W_0 *which sends every positive root to a negative root. (i.e. the longest element).*

Proof
(1) is clear since $q_{wb} = q_b$ for all $w \in W_0$.
(2) is easily checked, remembering that $q(s_a) = q_{a/2}q_a$ (3.1.6).
(3) follows from (1) and the definitions. □

The main result of this chapter is the following:

Theorem 4.1.2[1] *Let* $s\colon \Lambda \to \mathbb{C}^*$ *be a homomorphism such that* $s(t_b) \neq 1$ *for all* $b \in \Sigma_1$. *Then if* $z_0 \in Z^{++}$ *we have*

$$\omega_s(z_0^{-1}) = \frac{\delta(t_0)^{-1/2}}{P_{W_0}(q^{-1})} \sum_{w \in W_0} c(ws) \cdot (ws)(t_0)$$

where $t_0 = \nu(z_0)$ *and* $P_{W_0}(q^{-1}) = \sum_{w \in W_0} q(w)^{-1}$.

The proof will occupy sections 4.2–4.5. To see what the formula looks like in a particular case, take $G = \mathrm{SL}_{\ell+1}(k)$ as in (2.7.7). Here $\Sigma_1 = \Sigma_0$ and $q_a = q$ for all roots a, where q is the number of elements in the residue field of k. The roots are $e_i - e_j$ $(i \neq j)$ in the notation of (2.2.2) and we may write $s(t_{e_i - e_j}) = \xi_i \xi_j^{-1}$, where the ξ_i are non-zero complex numbers, determined up to a common factor by s. Then

$$c(s) = \prod_{i<j} \frac{1 - q^{-1}\xi_i^{-1}\xi_j}{1 - \xi_i^{-1}\xi_j} = \prod_{i<j} \frac{\xi_i - q^{-1}\xi_j}{\xi_i - \xi_j}.$$

Let ϖ be a generator of the maximal ideal of $\mathfrak{o}$, then we may take z_0 to be the diagonal matrix $z_0 = \mathrm{diag}(\varpi^{\lambda_0}, \ldots, \varpi^{\lambda_\ell})$ with $\lambda_0 \geq \lambda_1 \geq \cdots \geq \lambda_\ell$ (and $\sum \lambda_i = 0$). Then $s(t_0) = \xi_0^{\lambda_0} \ldots \xi_\ell^{\lambda_\ell}$; also by (3.2.11) $\delta(t)^{1/2} = q^{n(\lambda)}$ where

$$n(\lambda) = l\lambda_0 + (l-1)\lambda_1 + \cdots + \lambda_{\ell-1},$$

and

$$P_{W_0}(q^{-1}) = \prod_{i=1}^{l} \frac{1 - q^{-(i+1)}}{1 - q^{-1}}.$$

[1] The paper [Cas] provides an alternative proof of the result.

So we have

$$\omega_s(z_0^{-1}) = \frac{q^{-n(\lambda)}}{P_{W_0}(q^{-1})} \sum_{\sigma} \xi_{\sigma(0)}^{\lambda_0} \cdots \xi_{\sigma(l)}^{\lambda_\ell} \prod_{i<j} \frac{\xi_{\sigma(i)} - q^{-1}\xi_{\sigma(j)}}{\xi_{\sigma(i)} - \xi_{\sigma(j)}}$$

summed over all permutations σ of $\{0, 1, \ldots, l\}$.

4.2 Preliminary reductions

From (3.3.1) we have

$$\omega_s(g) = \int_K \phi_s(g^{-1}k)\mathrm{d}k \tag{4.2.1}$$

where $\phi_s(kzu^-) = (s\delta^{1/2}, \nu(z))$. *We shall prove* (4.1.2) *by computing this integral explicitly*. First we remark that ϕ_s is constant on each double coset KtU^-. For fixed $g \in G$, the set $g^{-1}K$ meets only finitely many of these double cosets (because they are open and mutually disjoint, and $g^{-1}K$ is compact), and therefore the integrand in (4.2.1) takes only finitely many values, each of which is of the form $(s\delta^{1/2}, t)$ for some $t \in \Lambda$. Hence $\omega_s(g)$ is of the form $s(\Phi_g)$ for some $\Phi_g \in \mathbb{C}[\Lambda]$ independent of s.

Next, since ω_s is constant on each double coset Kt we may assume that $g \in Z^{++}$. Write $g = z^{-1}$ and $t = \nu(z)$, so that $t^{-1} \in \Lambda^{++}$. Then

$$\omega_s(z^{-1}) = \int_K \phi_s(zk)\mathrm{d}k = \sum_{w \in W_0} \int_{BwB} \phi_s(zk)\mathrm{d}k. \tag{4.2.2}$$

Consider the integral $\int_{BwB} \phi_s(zk)\mathrm{d}k$. We have to look at $\phi_s(g)$ as g runs through $zBwB = zU_{(C_0)}wB = zU_{(C_0)}wBz^{-1}zwB = U_{t(C_0)}twB$ (using $zw = tw$.) Now $t^{-1}(0) \in \overline{C_0}$, hence $0 \in \overline{t(C_0)}$ and therefore $U_{t(C_0)} \subset K$. Hence the integration is effectively over the set twB, or equivalently (since $w \in K$) over $t'B$ where $t' = w^{-1}tw$.

Next, we have $B = U_{(C_0)}HU^-_{(C_0)}$ by (2.6.4), so that the integration is effectively over the set $z'U_{(C_0)}$, where z' is any element of $\nu^{-1}(t')$; and $z'U_{(C_0)} = U_{(C_0')}z'$ where $C_0' = t'(C_0)$ is a translate of the cone C_0. So we have

$$\int_{BwB} \phi_s(zk)\mathrm{d}k = \kappa \int_{U_{(C_0')}} \phi_s(u^+z')\mathrm{d}u^+ \tag{4.2.3}$$

where $\mathrm{d}u^+$ is the Haar measure on U^+, normalized as in §3.2, and κ is a constant independent of s. To evaluate κ we may put $s = \delta^{-1/2}$, so that ϕ_s is identically 1. Then the left-hand side of (4.2.3) is equal to the volume of BwB, which is $q(w)/P_{W_0}(q)$ in the notation introduced in §3.1, and the right-hand side is equal to

$$\kappa(U_{(C_0')} : U_{(C_0)}) = \kappa\delta(t'),$$

so that

$$\kappa = \frac{q(w)}{P_{W_0}(q)}\delta(t')^{-1}.$$

But also $\phi_s(u^+z') = \phi_s(u^+)(s\delta^{1/2}, t')$, and therefore (4.2.3) becomes

$$\int_{BwB} \phi_s(zk)\mathrm{d}k = \frac{q(w)}{P_{W_0}(q)}(s\delta^{-1/2}, t') \int_{U_{(C_0')}} \phi_s(u^+)\mathrm{d}u^+. \tag{4.2.4}$$

So if we put

$$J_w(s) = (w(s\delta^{-1/2}), t) \int_{U_{(C_0')}} \phi_s(u^+)\mathrm{d}u^+ \tag{4.2.5}$$

where $C_0' = t'(C_0) = w^{-1}tw(C_0)$, then from (4.2.2) and (4.2.4) we have

$$\omega_s(z^{-1}) = \frac{1}{P_{W_0}(q)} \sum_{w \in W_0} q(w)J_w(s). \tag{4.2.6}$$

Since $U_{(C_0')}$ is compact it intersects only finitely many of the double cosets KtU^-, and so as before we conclude that

$$J_w(s) = s(\Phi_{z,w}) \quad \text{for some } \Phi_{z,w} \in C_c(\Lambda). \tag{4.2.7}$$

To evaluate $\omega_s(z^{-1})$ we shall compute the integrals $J_w(s)$. First of all, if $w = 1$, then $t' = t$ and $\overline{C_0'} = t(\overline{C_0})$ contains the origin 0, so that $U_{(C_0')} \subset K$ and therefore $\phi_s(u^+) = 1$ for all $u^+ = U_{(C_0')}$. Consequently

$$J_1(s) = (s\delta^{1/2}, t) \tag{4.2.8}$$

(because the volume of $U_{(C_0')}$ is $\delta(t)$).

Now assume that $w \neq 1$. Then there exists a simple root $a \in \Pi_0$ such that $wa \in \Sigma_0^-$. The group $U_{(C_0)}$ is the semi-direct product of U_a and the group $U^a = \langle U_b : b \in \Sigma_0^+, b \neq a\rangle$. Hence $U_{(C_0')} = z'U_{(C_0)}z^{-1}$ is the semi-direct product of the group $V := z'U^a z'^{-1} \subset U^{(a)}$ where

$$U^{(a)} = \langle U_{(b)} : b \in \Sigma_0^+, b \neq a\rangle,$$

and the group $z'U_a z'^{-1} = U_{t'(a)}$. Hence

$$J_w(s) = (w(s\delta^{-1/2}), t) \int_V \int_{U_{t'(a)}} \phi_s(u^a u_a)\mathrm{d}u^a \mathrm{d}u_a \tag{4.2.9}$$

where $\mathrm{d}u^a$, $\mathrm{d}u_a$ are the Haar measures on $U^{(a)}$ and $U_{(a)}$ respectively, normalized as in §3.2.

In order to compute $J_w(s)$ we write $U_{t'(a)}$ as the difference between the sets $U_(a)$ and $U_(a) - U_{t'(a)}$, and consider the integrals

$$\begin{cases} J_w^1(s) = (w(s\delta^{-1/2}), t) \int_V \int_{U_{(a)}} \phi_s(u^a u_a)\mathrm{d}u^a \mathrm{d}u_a, \\ J_w^0(s) = (w(s\delta^{-1/2}), t) \int_V \int_{U_{(a)}-U_{t'(a)}} \phi_s(u^a u_a)\mathrm{d}u^a \mathrm{d}u_a. \end{cases} \tag{4.2.10}$$

Here the integration is no longer over a compact set, since neither $U_{(a)}$ nor $U_{(a)} - U_{t'(a)}$ is compact, and we shall therefore have to restrict s in order that the integrals shall converge.

4.3 The integral $\int_{U_{(a)}} \phi_s(u_a)\mathrm{d}u_a$

Proposition 4.3.1 *Let $a \in \Sigma_0^+$ and let $s \in \Lambda \to \mathbb{C}^*$ be such that $|s(t_a)| > 1$. Then the integral*

$$\int_{U_{(a)}} \phi_s(u_a)\mathrm{d}u_a$$

converges, and its value is equal to $c_0(a, s) = c(a/2, s)c(a, s)$ in the notation of §4.1.

Proof We split up $U_{(a)}$ into the union of U_a and the subsets

$$Y_r := U_{a-r} - U_{a-r+1}, \quad r \geq 1.$$

If $u_a \in U_a$, then $u_a \in K$ and therefore $\phi_s(u_a) = 1$. Consequently

$$\int_{U_{(a)}} \phi_s(u_a)\mathrm{d}u_a = 1.$$

Next consider $u_a \in Y_r$ $(r \geq 1)$. From (2.6.6) we can write $u_a = u_1 n u_2$ where $u_1, u_2 \in U_{-a+r}$ and $\nu(n) = w_{a-r}$. Hence

$$\phi_s(u_a) = \phi_s(u_1 n u_2) = \phi_s(n)$$

because $U_{-a+r} \subset U^- \cap K$. Choosing $n_a \in N$ such that $\nu(n_a) = s_a$ we have

$$\phi_s(u_a) = \phi_s(n_a n) = (s\delta^{1/2}, s_a \circ w_{a-r}) = (s\delta^{1/2}, t_a^{-r}).$$

Hence

$$\int_{Y_r} \phi_s(u_a)\mathrm{d}u_a = (s\delta^{1/2}, t_a^{-r}) \cdot \mathrm{vol}(Y_r)$$

and by (3.1.5) the volume of Y_r is equal to $q_{a/2}^{[r/2]} q_a^r - q_{a/2}^{[(r-1)/2]} q_a^{r-1}$. Consequently the value of the integral is

$$1+\sum_{r=1}^{\infty}\frac{(q_{a/2}^{[r/2]}q_a^r - q_{a/2}^{[(r-1)/2]}q_a^{r-1})}{s(t_a)^r q_{a/2}^{r/2}q_a^r},$$

making use of $\delta(t_a) = q_{a/2}q_a^2$ (3.2.7). This is a geometric series[2] with common ratio $s(t_a)^{-2}$, hence converges if and only if $|s(t_a)| > 1$. Summing the series, we get the value announced in (4.3.1). □

4.4 Reduction of $J_w(s)$

Consider first the integral $J_w^1(s)$ defined in (4.2.10). We recall that in the definition of $J_w(s)$, a is a simple root such that $wa \in \Sigma_0^-$.

Proposition 4.4.1 *Suppose that $|s(t_a)| > 1$. Then the integral $J_w^1(s)$ converges and its value is $c_0(a,s)J_{w'}(s_a s)$ where $w' = ws_a$.*

Proof Choose $n_a \in N$ such that $\nu(n_a) = s_a$. Since $n_a \in K$ we have $\phi_s(u^a u_a) = \phi_s(n_a^{-1}u^a n_a \cdot n_a^{-1}u_a)$. Write $n_a^{-1}u^a n_a$ in the form kz_1u^-, where $k \in K, z_1 \in Z, u^- \in U^-$. Then

$$\phi_s(u^a u_a) = \phi_s(z_1 u^- \cdot n_a^{-1}u_a) = \phi_s(z_2 \cdot n_a u^- n_a^{-1} \cdot u_a)$$

where $z_2 = n_a z_1 n_a^{-1}$, and $n_a u^- n_a^{-1} \in n_a U^- n_a^{-1} \in U^{(-a)} \cdot U_{(a)}$. Hence $n_a u^- n_a^{-1}$ can be written uniquely in the form $u^{-a}u'_a$ $(u^{-a} \in U^{(-a)}, u'_a \in U_{(a)})$ and therefore

$$\phi_s(u^a u_a) = \phi_s(z_2 u^{-a}u'_a u_a) = \phi_s(z_2 u'_a u_a)$$

since $U_{(a)}$ normalizes $U^{(-a)}$ and ϕ_s is invariant on the right with respect to $U^- \supset U^{(-a)}$. So, putting $u''_a = u'_a u_a$, we have

$$\phi_s(u^a u_a) = \phi_s(z_2 u''_a) = \phi_s(z_2 u''_a z_2^{-1})(s\delta^{1/2}, \nu(z_2))$$

and

$$(s\delta^{1/2}, \nu(z_2)) = (s_a s, \nu(z_1))(\delta^{1/2}, \nu(z_2)) = \phi_{s_a s}(n_a^{-1}u^a n_a) \cdot (\delta^{1/2}, \nu(z_1^{-1}z_2)).$$

Hence

$$\phi_s(u^a u_a) = \phi_s(z_2 u''_a z_2^{-1}) \cdot \phi_{s_a s}(n_a^{-1}u^a n_a) \cdot (\delta^{1/2}, \nu(z_1^{-1}z_2)).$$

[2] It is convenient to consider the cases r even and odd separately.

As u^a runs through the subgroup $V = z'U^a z'^{-1}$ defined in §4.2, $n_a^{-1}u^a n_a$ runs through $V' = n_a^{-1}Vn_a$ and so we have

$$J_w^1(s) = (w(s\delta^{-1/2}), t) \int_{V'} \phi_{s_a s}(u^a)(\delta^{1/2}, \nu(z_1^{-1}z_2))\mathrm{d}u^a$$
$$\times \int_{U_{(a)}} \phi_s(z_2 u_a'' z_2^{-1})\mathrm{d}u_a.$$

But

$$\int_{U_{(a)}} \phi_s(z_2 u_a'' z_2^{-1})\mathrm{d}u_a = \int_{U_{(a)}} \phi_s(z_2 u_a z_2^{-1})\mathrm{d}u_a = \Delta_a(z_2)^{-1} \int_{U_{(a)}} \phi_s(u_a)\mathrm{d}u_a$$

and by (4.3.1) this integral converges, since $|s(t_a)| > 1$. Since V' is compact it follows that $J_w^1(s)$ converges, and we have

$$J_w^1(s) = (w(s\delta^{-1/2}), t)c_0(a, s) \int_{V'} \phi_{s_a s}(u^a)\mathrm{d}u^a \qquad (4.4.2)$$

since by (3.2.9)

$$(\delta^{1/2}, \nu(z_1^{-1}z_2)) = \delta_a(\nu(z_2)) = \Delta_a(z_2).$$

To complete the proof of (4.4.1) we shall show that

Lemma 4.4.3 *For any* $s \in \mathrm{Hom}(\Lambda, \mathbb{C}^\times)$,

$$J_{w'}(s) = (w's, t)(w\delta^{-1/2}, t) \int_{V'} \phi_s(u^a)\mathrm{d}u^a.$$

Proof of the lemma From the definition (4.2.5),

$$J_{w'}(s) = (w'(s\delta^{-1/2}, t) \int_{U_{(C_0'')}} \phi_s(u^a)\mathrm{d}u^a$$

where $C_0'' = (w'^{-1}tw')(C_0)$, so that $U_{(C_0'')}$ is the semi-direct product of V' and U_β, where $\beta = (w'^{-1}tw)a = a - w'a(x)$ where $x = t(0)$.

Since $w' = ws_a$, we have $\beta(0) = a(0) + wa(x) = wa(x)$ which is ≥ 0 because wa is a negative root, and $x \in -\overline{C_0}$. Hence $U \subset K$, so that

$$\begin{aligned}\int_{U_\beta} \phi_s(u_a)\mathrm{d}u_a &= \mathrm{vol}(U_\beta)\\ &= \delta_a(w'^{-1}tw')\\ &= \delta(w'^{-1}tw')^{1/2}\delta(w^{-1}tw)^{-1/2} \quad \text{by (3.2.9)}\\ &= (w'\delta^{1/2}, t)\cdot(w\delta^{-1/2}, t).\end{aligned}$$

This completes the proof of (4.4.3) and hence also of (4.4.1). □

Next we consider the integral

$$J_w^0(s) = (w(s\delta^{-1/2}), t) \int_V \int_{U(a)-U_\alpha} \phi_s(u^a u_a) \mathrm{d}u^a \mathrm{d}u_a$$

where $\alpha = (w^{-1}tw)(a) = a - wa(x)$ if $x = t(0)$.

Proposition 4.4.4 *Suppose that $|s(t_a)| > 1$. Then the integral $J_w^0(s)$ converges and is equal to*

$$(c_0(a,s) - 1)J_{w'}(s)$$

where as before $w' = ws_a$.

Proof As in §4.3 put $Y_r = U_{a-r} - U_{a-r+1}$ so that $U_{(a)} - U_\alpha$ is the disjoint union of the sets Y_r for $r \geq 1 + wa(x) > 0$. Consider the integral

$$\int_V \int_{Y_r} \phi_s(u^a u_a) \mathrm{d}u^a \mathrm{d}u_a.$$

The calculation proceeds on similar lines to those of §4.3. By (2.6.6) we may write $u_a = u_1 n u_2$ with $u_1, u_2 \in U_{-a+r}$ and $\nu(n) = w_{a-r}$. Hence

$$\begin{aligned} \phi_s(u^a u_a) &= \phi_s(u^a u_1 n) \quad (\text{because } u_2 \in U^-) \\ &= \phi_s(u_1^{-1} u^a u_1 n_a^{-1} \cdot n_a n) \quad (\text{because } u_1 \in K) \ (\text{where } n_a \in \nu^{-1}(s_a)) \\ &= \phi_s(u_1^{-1} u^a u_1 n_a^{-1}) \cdot (s\delta^{1/2}, t_a^{-r}) \end{aligned}$$

because $n_a n \in Z$ and $\nu(n_a n) = s_a w_{a-r} = t_a^{-r}$.

Hence

$$\int_V \int_{Y_r} \phi_s(u^a u_a) \mathrm{d}u^a \mathrm{d}u_a = \left(\int_V \phi_s(u_1^{-1} u^a u_1 n_a^{-1}) \mathrm{d}u^a \right) \times (s\delta^{1/2}, t_a^{-r}) \mathrm{vol}(Y_r).$$

But u_1 normalizes V and does not change the measure, so that

$$\begin{aligned} \int_V \phi_s(u_1^{-1} u^a u_1 n_a^{-1}) \mathrm{d}u^a &= \int_V \phi_s(u^a n_a^{-1}) \mathrm{d}u^a \\ &= \int_{V'} \phi_s(u^a) \mathrm{d}u^a \end{aligned}$$

where as before $V' = n_a^{-1} V n_a$. Consequently

$$\begin{aligned} &\int_V \int_{U(a)-U_\alpha} \phi_s(u^a u_a) \mathrm{d}u^a \mathrm{d}u_a \\ &= \left(\int_{V'} \phi_s(u^a) \mathrm{d}u^a \right) \sum_{r=1+wa(x)}^{\infty} (s\delta^{1/2}, t_a)^{-r} (q_{a/2}^{[r/2]} q_a^r - q_{a/2}^{[(r-1)/2]} q_a^{r-1}). \end{aligned}$$

summing this geometric series, which converges if and only if $|s(t_a)| > 1$, and using (4.4.3) now leads to the result. [Recall that $wa(x)$ is an even integer if $q_{a/2} \neq 1$, by (3.1.4).] □

These calculations lead to the following *reduction formula*:

Theorem 4.4.5 *If $s(t_a) \neq 1$, then*

$$J_w(s) = c_0(a,s)J_{w'}(s_a s) - (c_0(a,s) - 1)J_{w'}(s),$$

where $a \in \Pi_0$ is such that wa is negative, and $w' = ws_a$.

Proof If $|s(t_a)| > 1$, this follows directly from (4.4.1) and (4.4.4). By (4.2.7) the $J's$ are polynomials in $\xi_i = s(t_{a_i})$ and ξ_i^{-1} $(1 \leq i \leq l)$, where $a_1, \ldots, a_\ell$ are the simple roots. Hence it is clear (e.g. by analytic continuation) that (4.4.5) is valid whenever the right-hand side is defined, i.e. whenever $s(t_a) \neq 1$. □

This reduction formula is the key result. In the next section we shall use it first to prove that $\omega_s = \omega_{ws}$ for all $w \in W_0$, which was postponed from Chapter III (3.3.3), and then to complete the proof of the formula (4.1.2).

4.5 End of the proof

We shall begin by proving (3.3.3). Clearly it is enough to show that $\omega_s(z^{-1}) = \omega_{s_a s}(z^{-1})$ for all $a \in \Pi_0$, and we may assume that $s(t_a) \neq 1$, otherwise $s = s_a s$ and there is nothing to prove.

Put $I_w(s) = J_w(s) - J_w(s_a s)$ and suppose first that wa is a negative root, and $a \in \Pi_0$. Then by (4.4.5)

$$J_w(s) = -c_0(a,s)I_{w'}(s) + J_{w'}(s)$$

where $w' = ws_a$. Replacing s by $s_a s$ and subtracting, we get

$$I_w(s) = (1 - c_0(a,s) - c_0(-a,s))I_{w'}(s) = -q(s_a)^{-1}I_{w'}(s)$$

by (4.1.1). Since wa is negative, there is a reduced word for w ending with s_a, and therefore $q(w) = q(w')q(s_a)$ by (3.1.7). Hence we have

$$q(w)I_w(s) = -q(w')I_{w'}(s). \tag{4.5.1}$$

If on the other hand wa is positive, then the roles of w' and w are interchanged, and (4.5.1) is therefore true for all $w \in W_0$. Summing over $w \in W_0$, it follows that

$$\sum_{q \in W_0} q(w)I_w(s) = 0$$

and hence $w_s(z^{-1}) = \omega_{s_a s}(z^{-1})$ by (4.2.6).

Let $w \in W_0$ and let $w = w_1 \cdots w_r$ be a reduced word in the generators s_a, $a \in \Pi_0$. Let E be the set of all products $w_1^{\epsilon_1} \cdots w_r^{\epsilon_r}$ where ϵ_i is 0 or 1.

If $s(t_b) \neq 1$ for all $b \in \Sigma_1$, we shall say that s is *nonsingular*.

Lemma 4.5.2 *Suppose s is nonsingular. Then*

$$J_w(s) = \delta(a)^{1/2} \sum_{w' \in E} \lambda_{w,w'}(s) \cdot (w's, t)$$

with coefficients $\lambda_{w,w'}$ independent of t. Also

$$\lambda_{w,w'}(s) = \prod_a c_0(a, s)$$

the product being taken over the roots $a \in \Sigma_0^+$ such that wa is negative.

Proof By induction on the length of $\ell(w)$. If $\ell(w) = 0$ then $w = 1$ and $J_1(s) = \delta^{1/2}(s, t)$ by (4.2.8). If $\ell(w) > 0$, apply (4.4.5) with $w' = w_1 \dots w_{r-1}$. □

From (4.5.2) and (4.2.6) it follows that we have

$$\omega_s(z^{-1}) = \delta(t)^{1/2} \sum_{w \in W_0} \mu_w(s)(ws, t) \tag{4.5.3}$$

with coefficients $\mu_w(s)$ independent of $t = \nu(z)$. Consider in particular the coefficient $\mu_{w_0}(s)$, where w_0 is the unique element of W_0 which sends every positive root to a negative root. This coefficient comes only from $J_{w_0}(s)$ and hence by (4.5.2) we have

$$\mu_{w_0}(s) = \frac{q(w_0)}{P_{W_0}(q)} \lambda_{w_0,w_0}(s) = \frac{c(s)}{P_{W_0}(q^{-1})} \tag{4.5.4}$$

since $q(w_0)^{-1} P_{W_0}(q) = P_{W_0}(q^{-1})$.

Next we have

Proposition 4.5.5 *Assume that $|s|$ is nonsingular (i.e. that $|s(t_a)| \neq 1$ for all $a \in \Sigma_0$). Then the homomorphisms ws with $w \in W_0$ are linearly independent over $\mathbb{C}$.*

Assuming this for the moment, since $\omega_s = \omega_{w's}$ for all $w' \in W_0$ we have from (4.5.3)

$$\sum_{w \in W_0} \mu_w(s)(ws, t) = \sum_{w \in W_0} \mu_w(w's)(ww's, t)$$

and therefore (assuming s nonsingular) by (4.5.5) $\mu_w(s) = \mu_1(ws)$ for all $w \in W_0$. In particular, therefore $\mu_1(w_0 s) = c(s)/P_{W_0}(q^{-1})$ by (4.5.4); or, replacing s by $w_0 s$,

$$\mu_1(s) = \frac{c(s^{-1})}{P_{W_0}(q^{-1})}$$

by (4.1.1). Hence (4.5.3) now becomes

$$\omega_s(z^{-1}) = \frac{\delta(t)^{1/2}}{P_{W_0}(q^{-1})} \sum_{w \in W_0} c(ws^{-1})(ws, t) \tag{4.5.6}$$

provided that $|s|$ is nonsingular. However, this restriction is superfluous, for the left-hand side of (4.5.6) is of the form $s(\Phi_t)$ for some $\Phi_t \in \mathbb{C}[\Lambda]$, and the right-hand side is of the form $s(\Psi_t)$ for some $\Psi_t \in \mathbb{C}(\Lambda)$, the field of fractions of $\mathbb{C}[\Lambda]$. Since Φ_t and Ψ_t take the same values on a dense open set in $S = \mathrm{Hom}(\Lambda, \mathbb{C}^*)$, it follows that they are equal and therefore (4.5.6) is valid whenever the right-hand side is defined, i.e. provided s is nonsingular.

Finally, to derive the formula (4.1.2) we have merely to replace s by s^{-1} and put $z_0 = z^{-1}$ in (4.5.6), bearing in mind (3.3.2).

There remains the proof of (4.5.5). This will follow from

Lemma 4.5.7 *Let $s_1, \dots, s_n$ $(n \geq 1)$ be distinct elements of $Hom(\Lambda, \mathbb{C}^\times)$. Then they are linearly independent.*

Proof The argument is a familiar one. If the s_i are linearly independent, choose a linear relation between them containing as few non-zero terms as possible, say $\sum_{i=1}^r \lambda_i s_i = 0$ (renumbering the s_i if necessary) with each $\lambda_i \neq 0$. Since $s_1 \neq s_r$ there exists $t_0 \in \Lambda$ such that $s_1(t_0) \neq s_r(t_0)$. Then we have

$$\sum_{i=1}^r \lambda_i s_i(t) = 0, \quad \sum_{i=1}^r \lambda_i s_i(t_0) s_i(t) = 0$$

for all $t \in \Lambda$, so by subtraction we get

$$\sum_{i=1}^r \lambda_i \left(s_i(t_0) - s_r(t_0)\right) s_i = 0,$$

contradicting the minimality of r. □

From (4.5.7) it follows that if the set $\{ws : w \in W_0\}$ is linearly dependent, then $w_1 s = w_2 s$ for some $w_1 \neq w_2$ in W_0, and hence s is fixed by an element $\neq 1$ in W_0. Hence the same is true of $\log|s| \in \mathrm{Hom}(\Lambda, \mathbb{R})$. Identifying $\log|s|$ with an element $u \in V^*$ as in (3.3.13), it follows that u has non-trivial stabilizer under the action of W_0 on V^* and hence lies on a reflecting hyperplane. In other words, we have $u(a^\vee) = 0$ for some $a \in \Sigma_0$, and this means that $|s(t_a)| = 1$. This proves (4.5.5) and hence completes the proof of (4.1.2).

Remark 4.5.8 If $s = \delta^{-1/2}$ (which is nonsingular) then ω_s is identically 1.[3] Hence also $\omega_{\delta^{1/2}} = 1$ by (3.3.2). Putting $s = \delta^{1/2}$ in (4.1.2) gives therefore

$$P_{W_0}(q^{-1}) = \delta(t_0)^{-1/2} \sum_{w \in W_0} c(w\delta^{1/2})(w\delta^{1/2}, t_0)$$

[3] From this observation we can determine the Satake parameter for the trivial representation, which is obviously K-spherical.

for all $t_0 \in \Lambda^{++}$. Consider $c(w\delta^{1/2})$ for $w \neq 1$. There exists a simple root a such that $w^{-1}a$ is negative. Let $b = -w^{-1}a$, then we have $t_b^{-1} = wt_a w^{-1}$ and therefore

$$w\delta^{1/2}(t_b)^{-1} = \delta^{1/2}(t_a) = q_{a/2}^{1/2} q_a = q_{b/2}^{1/2} q_b$$

so that $c(b, w\delta^{1/2}) = 0$ and therefore $c(w\delta^{1/2}) = 0$. Consequently the sum above reduces to the term corresponding to $w = 1$, that is

$$P_{W_0}(q^{-1}) = c(\delta^{1/2}) = \prod_{b \in \Sigma_1^+} \frac{1 - q_{b/2}^{-1/2} q_b^{-1} \delta(t_b)^{-1/2}}{1 - q_{b/2}^{-1/2} \delta(t_b)^{-1/2}}.$$

This suggests (but does not prove) the following multiplicative formula for the Poincaré polynomial $P_{W_0}(\xi)$ of the root system Σ_1. For $b \in \Sigma_1$, define

$$\eta_b = \prod_{c \in \Sigma_1^+} \xi_c^{c(b^\vee)}.$$

Then

$$P_{W_0}(\xi) = \prod_{b \in \Sigma_1^+} \frac{1 - \xi_{b/2}^{1/2} \xi_b \eta_b^{1/2}}{1 - \xi_{b/2}^{1/2} \eta_b^{1/2}} \tag{4.5.9}$$

with the convention that $\xi_{b/2} = 1$ if $b/2 \notin \Sigma_1$. In fact (4.5.9) is true and can be proved by elementary methods. Putting $\xi_b = \xi$ for all $b \in \Sigma_1$, and assuming that Σ_1 is reduced (i.e. that $\Sigma_1 = \Sigma_0$), (4.5.9) specializes to give

$$\sum_{w \in W_0} \xi^{\ell(w)} = \prod_{a \in \Sigma_0^+} \frac{1 - \xi^{1+\mathrm{ht}(a^\vee)}}{1 - \xi^{\mathrm{ht}(a^\vee)}} = \prod_{a \in \Sigma_0^+} \frac{1 - \xi^{1+\mathrm{ht}(a)}}{1 - \xi^{\mathrm{ht}(a)}}$$

(since Σ_0 and $\Sigma_0^\vee$ have the same Weyl group W_0).

4.6 The singular case

Next we consider the problem of finding an expression for $\omega_s(z^{-1})$ when s is singular, i.e. when $s(t_b) = 1$ for some $b \in \Sigma_1$. We have remarked earlier that $\omega_s(z^{-1})$ is of the form $s(\Psi_t)$ (where $t = v(z)$) for some polynomial $\Psi_t \in \mathbb{C}[\Lambda]_0^W$, independent of s. Hence we can calculate $\omega_s(z^{-1})$ for singular s as the limit of $\omega_{s_1}(z^{-1})$ as $s_1 \to s$ through nonsingular values of s_1. *This limit can be computed by applying L'Hôpital's rule in exactly the same way that the degree of an irreducible representation of a compact Lie group is derived from Weyl's character formula.*

For simplicity we shall consider first the "most singular" case, namely $s = 1$. Then we have to evaluate the limit as $s \to 1$ of

$$\sum_{w \in W_0} c(ws)(ws, t) \quad (t \in \Lambda^{++}).$$

In calculating the limit, we may make $s \to 1$ through real values. So let s be real and let σ be the linear form on A defined by $\sigma(t(0)) = \log s(t)$ for all $t \in \Lambda$. Then $s(t_a) = e^{\sigma(a^\vee)}$, so that

$$c(s) = \prod_{a \in \Sigma_0^+} \frac{(1 + q_{a/2}^{-1/2} e^{-\sigma(2a^\vee)})(1 - q_{a/2}^{-1/2} q_a^{-1} e^{-\sigma(2a^\vee)})}{1 - e^{-\sigma(2a^\vee)}}.$$

Let us temporarily write

$$\prod_{a \in \Sigma_0^+} (1 + q_{a/2}^{-1/2} e^{-\sigma(2a^\vee)})(1 - q_{a/2}^{-1/2} q_a^{-1} e^{-\sigma(2a^\vee)}) = \sum_{y \in L^\vee} A_y e^{-\sigma(y)}$$

where $L^\vee$ is the lattice in A spanned by the $a^\vee$ for $a \in \Sigma_0$, and all but finitely many of the coefficients A_y are zero. Then

$$c(s)(s, t) = \frac{\sum A_y \exp(\sigma(x + p - y))}{\sum_{w \in W_0} \epsilon(w) \exp(w\sigma(p))}$$

where $x = t(0)$ and $p = \sigma_{a \in \Sigma_0^+} a^\vee$, and $\epsilon(w)$ is the signature (± 1) of $w \in W_0$. So we have

$$c(ws)(ws, t) = \sum_y A_y \frac{\sum_{w \in W_0} \epsilon(w) \exp(w\sigma(x + p - y))}{\sum_{w \in W_0} \epsilon(w) \exp(w\sigma(p))}$$

and therefore the limit of the left-hand side as $s \to 1$, that is as $\sigma \to 0$, is

$$\sum_y A_y \frac{\pi(x + p - y)}{\pi(p)}$$

where π is the polynomial function $\prod_{a \in \Sigma_0^+} a$ on A.

To express this result in a concise form let us make the following convention: if $F = \sum \lambda_i t_i \in \mathbb{C}[\Lambda]$, where $\lambda_i \in \mathbb{C}$ and $t_i \in \Lambda$, then $\pi \circ F$ denotes the polynomial function $\sum \lambda_i \pi \circ t_i$ on A: that is,

$$(\pi \circ F)(x) = \sum \lambda_i \pi(t_i(x)), \quad (x \in A).$$

Then we have proved

Proposition 4.6.1 *Let* $t = \nu(z) \in \Lambda^{++}$ *and put* $x = t(0)$. *Let*

$$F = \prod_{a \in \Sigma_0^+} (1 + q_{a/2}^{-1/2} t_a^{-1})(1 - q_{a/2}^{-1/2} q_a^{-1} t_a^{-1}) \in \mathbb{C}[\Lambda].$$

Then

$$\omega_1(z^{-1}) = \frac{\delta(t)^{-1/2}}{P_{W_0}(q^{-1})} \frac{(\pi \circ F)(x + p)}{\pi(p)}$$

where $\pi = \prod_{a \in \Sigma_0^+} a$ *and* $p = \prod_{a \in \Sigma_0^+} a^\vee$.

Hence $\omega_1(z^{-1})$ is of the form $\Phi(x)e^{-\lambda(x)}$, where Φ is a polynomial and λ is a linear form on A such that $\lambda(x) > 0$ for all $x \in \overline{C_0}$ (except $x = 0$). If follows that $\omega_1(z^{-1}) \to 0$ as $x \to \infty$ in any direction in the cone $\overline{C_0}$ and consequently that the spherical function ω_1 is *bounded*.

The case of arbitrary singular s is quite analogous, and we shall merely state the result. Let Σ_{0s} be the subsystem of Σ_0 consisting of all roots a such that $|s(t_a)| = 1$; put

$$c^s(s) = \prod_{a \in \Sigma_0^+ - \Sigma_{0s}^+} c_0(a, s)$$

$$F_s = \prod_{a \in \Sigma_{0s}^+} (1 + q_{a/2}^{-1/2} t_a^{-1})(1 - q_{a/2}^{-1/2} q_a^{-1} t_a^{-1})$$

$$\pi_s = \prod_{a \in \Sigma_{0s}^+} a, \quad p_s = \prod_{a \in \Sigma_{0s}^+} a^\vee.$$

Proposition 4.6.2 *We have, for all* $s_0 \in S = \mathrm{Hom}(\Lambda, \mathbb{C}^\times)$, *and* $z \in Z^{++}$,

$$\omega_{s_0}(z^{-1}) = \frac{\delta(t)^{-1/2}}{P_{W_0}(q^{-1})} \sum_s \frac{(\pi \circ F_s)(x + p_s)}{\pi_s(p_s)} c^s(s)(s, t)$$

summed over all s *in the orbit* $W_0 s_0$ *of* s_0. *As before,* $t = \nu(z)$ *and* $x = t(0)$. *Hence* $\omega_{s_0}(z^{-1})$ *is of the form*

$$\omega_{s_0}(z^{-1}) = \sum_s \Phi_s(x)(s\delta^{-1/2}, t) \tag{4.6.3}$$

where the Φ_s *are polynomial functions on* A, *depending on* s.

4.7 Bounded spherical functions

Let $s_0 \in S = \mathrm{Hom}(\Lambda, \mathbb{C}^\times)$. As in (3.3.13) we shall identify $\log|s_0| \in \mathrm{Hom}(\Lambda, \mathbb{R})$ with an element of V^*, namely the linear form on A whose value at $t(0)$ is $\log|s_0(t)|$ for all $t \in \Lambda$. In particular, with this identification it makes sense to talk of the convex hull in V^* of the set

$$D = \{\log w\delta^{1/2} : w \in W_0\}.$$

Theorem 4.7.1 *The spherical function ω_{s_0} is bounded* $\Leftrightarrow$ $\log|s_0|$ *lies in the convex hull of D.*

Proof Let $C_0^*, C_0^{*\perp}$ be the images in V^* of the cones $C_0, C_0^\perp$ respectively, under the isomorphism of A onto V^* defined by the scalar product. Then $\overline{C_0^*}$, the closure of C_0^*, is a fundamental region for the action of W_0 on V^*.

Since $\omega_{s_0} = \omega_{ws_0}$ for all $w \in W_0$, we may assume that $\log|s_0| \in \overline{C_0^*}$, or equivalently that $|s(t_a)| \geq 1$ for all $a \in \Sigma_0^+$. It is not hard to show that

$$D \cap \overline{C_0^*} = (\log \delta^{1/2} - C_0^{*\perp}) \cap \overline{C_0^*}$$

so that we have to prove that

$$\omega_{s_0} \text{ is bounded} \Leftrightarrow \log \delta^{1/2} - \log|s_0| \in C_0^{*\perp}$$

or equivalently that

$$\omega_{s_0} \text{ is bounded} \Leftrightarrow |(\delta^{-1/2}s_0, t)| \leq 1, \forall t \in \Lambda^{++}.$$

Suppose first that ω_{s_0} is bounded. Let $z \in Z^{++}$ $t = \nu(z)$, $x = t(0)$ and assume that $x \in C_0$, i.e. that x does not lie on a wall of the chamber C_0. By (4.6.3) we have

$$\begin{aligned}\omega_{s_0}(z^{-n}) &= \sum_{s \in W_0 s_0} \Phi_s(nx)(\delta^{-1/2}s, t^n) \\ &= (\delta^{-1/2}s_0, t^n)\left\{\Phi_{s_0}(nx) + \sum_{s \neq s_0} \Phi_s(nx)\frac{(s, t^n)}{(s_0, t^n)}\right\}.\end{aligned}$$

Consider the terms in this sum. We have $s = ws_0 \neq s_0$, hence

$$\left|\frac{(s, t^n)}{(s_0, t^n)}\right| = \left|\frac{(ws_0, t^n)}{(s_0, t^n)}\right| = e^{-n\lambda(x)} \text{ say,}$$

where $\lambda = \log|s_0| - w\log|s_0| \in C_0^{*\perp}$. Since $x \in C_0$ it follows that $\lambda(x) > 0$. Also $\Phi_s(nx)$, for fixed x, is a polynomial in n. Hence, as $n \to \infty$, we have

$$\left|\Phi_s(nx)\frac{(s, t^n)}{(s_0, t^n)}\right| = |\Phi_s(nx)|e^{-n\lambda(x)} \to 0$$

and therefore

$$\omega_{s_0}(z^{-n}) \sim (\delta^{-1/2}s_0, t^n)\, \Phi_{s_0}(nx).$$

The polynomial Φ_{s_0} is not identically zero. Since ω_{s_0} is bounded it follows that, *for almost all* $t \in \Lambda^{++}$, we have

$$|(\delta^{-1/2}s_0, t)| \le 1.$$

It follows then that $|(\delta^{-1/2}s_0, t)| \le 1$ for all $t \in \Lambda^{++}$.

Conversely, suppose that s_0 satisfies this condition. Since the set of bounded spherical functions is closed in the set Ω of all spherical functions (see e.g. Tamagawa [16, p. 378]), we may assume that s_0 is nonsingular. But then we can use the original formula (4.1.2):

$$\omega_{s_0}(z^{-1}) = \frac{(\delta^{-1/2}s_0, t)}{P_{W_0}(q^{-1})}\left\{c(s_0) + \sum_{w\neq 1}\frac{(ws_0, t)}{(s_0, t)}\right\}.$$

Since $s_0 \in C_0^*$ and $t \in \Lambda^{++}$ we have

$$|(ws_0, t)| \le |(s_0, t)|$$

for all $w \neq 1$, and therefore

$$|\omega_{s_0}(z^{-1})| \le \frac{1}{P_{W_0}(q^{-1})}\sum_{w\in W_0}|c(ws_0)|.$$

Consequently ω_{s_0} is bounded. □

4.8 Notes on Chapter IV

The formula (4.1.2) for the case of a Chevalley group was announced in [12]. In its present form it dates from 1968 but appears here for the first time. (4.7.1) is strictly analogous to the real case (see [11]).

4.9 Additions to Chapter IV

4.9.1 The spherical Plancherel formula for the extended affine Hecke algebra

The Hecke algebra $\mathcal{H} := \mathcal{H}(W, \mathbf{q})$ from §2.9.2 becomes an involutive algebra with respect to

$$\left(\sum_{w\in W} b_w T_w\right)^* := \sum_{w\in W} \overline{b}_{w^{-1}} T_w, \quad \text{with } b_w \in \mathbb{C}. \tag{4.9.1}$$

The canonical trace $\tau\colon \mathcal{H} \to \mathbb{C}$ is defined by

$$\tau\left(\sum_{w\in W} b_w T_w\right) := b_0,$$

where $b \mapsto \overline{b}$ is the complex conjugation.

It gives rise to the canonical hermitian inner product

$$(f_1, f_2) := \tau(f_1 f_2^*) = \sum_{w\in W} b_{1,w}\overline{b}_{2,w} q^{\ell(w)},$$

where $f_i = \sum_{w\in W} b_{i,w} T_w \in \mathcal{H}$, with $i = 1, 2$.

We denote by $\mathcal{T}$ the complex algebraic torus $\mathcal{T} := \mathrm{Hom}(Y, \mathbb{C}^\times)$ of complex characters of the lattice Y, and set

$$\nu(t) := (c(t)c(t^{-1}))^{-1} = \prod_{a\in\Sigma_{0,+}} (c_0(a,t)c_0(a,t^{-1}))^{-1}. \tag{4.9.2}$$

By [Op4, Corollary A.12], for every point $t_0 \in \mathcal{T}$, the pole order of $\nu(t)$ at t_0 is at most equal to the rank of Σ_0.

Definition An element t_0 of $\mathcal{T}$ is called a *residual point* if the pole order of $\nu(t)$ at t_0 is equal to the rank of Y. The set of residual points in $\mathcal{T}$ is a finite union of W_0-orbits, and this set is nonempty only if the rank of Σ_0 is equal to that of Y.

Recall that $\mathrm{Z}(\mathcal{H})$ denotes the centre of $\mathcal{H}$ (see (2.9.14)). The space of complex homomorphisms of $\mathrm{Z}(\mathcal{H})$ is canonically isomorphic to the (categorical) quotient $W_0\backslash\mathcal{T}$. Let $\mathrm{Irr}(\mathcal{H})$ denote the set of equivalence classes of irreducible representations of $\mathcal{H}$. There is a continuous, finite, surjective map (called the *algebraic central character*)

$$z\colon \mathrm{Irr}(\mathcal{H}) \to W_0\backslash\mathcal{T}. \tag{4.9.3}$$

We define $L^2(\mathcal{H})$ to be the Hilbert space completion of $\mathcal{H}$, and we denote by $B(L^2(\mathcal{H})$ the algebra of bounded linear operators on $L^2(\mathcal{H})$. The regular representation λ of $\mathcal{H}$ extends to a representation of $\mathcal{H}$ in $B(L^2(\mathcal{H}))$. This gives $\mathcal{H}$ the structure of a unital Hilbert algebra.

Definition The reduced C^*-algebra $C_{\mathrm{r}}^*(\mathcal{H})$ of $\mathcal{H}$ is the norm closure of $\lambda(\mathcal{H})$ in $B(L^2(\mathcal{H})$. We identify $C_{\mathrm{r}}^*(\mathcal{H})$ with a dense subspace of $L^2(\mathcal{H})$ via the continuous injection $x \mapsto x(1)$.

Definition An irreducible representation (π, V) of $\mathcal{H}$ is a *discrete series representation* if it is equivalent to a subrepresentation of $(L^2(\mathcal{H}), \lambda)$, or equivalently, if the character of π extends continuously to $L^2(\mathcal{H})$.

Let $\widehat{\mathcal{H}}$ denote the *unitary dual* of $\mathcal{H}$, that is, the set of equivalence classes of irreducible unitary representations of $\mathcal{H}$. We define $C(\mathcal{H})$ to be the formal completion of $\mathcal{H}$ with respect to the basis $(T_w)_{w\in W}$. Hence $f \in C(\mathcal{H})$ means that $f = \sum b_w T_w$

is a formal infinite sum with complex coefficients. For $f \in C(\mathcal{H})$ and $f' \in \mathcal{H}$, the product $ff' \in C(\mathcal{H})$ and the hermitian pairing (f, f') are well defined.

There exists a unique positive Borel measure μ_{Pl} on $C_{\mathrm{r}}^*(\mathcal{H})$, the canonical *Plancherel measure of* $\mathcal{H}$ (see [Dix]), such that we have the following decomposition of h in irreducible characters of $C_{\mathrm{r}}^*(\mathcal{H})$:

$$h = \int_{\pi \in \mathrm{Irr}(C_{\mathrm{r}}^*(\mathcal{H})} \theta_\pi \, \mathrm{d}\mu_{\mathrm{Pl}}(\pi). \tag{4.9.4}$$

The c-function allows us to obtain the following algebraic criterion for a central character $W_0 t \in W_0 \backslash \mathcal{T}$ to be the central character of a discrete series representation of $\mathcal{H}$: an orbit $W_0 r \in W_0 \backslash \mathcal{T}$ is the algebraic central character of a discrete series representation of $\mathcal{H}$ if and only if $r \in \mathcal{T}$ is a residual point (see [Op4, Theorem 3.29]). In particular, the set of equivalence classes of discrete series representations of $\mathcal{H}$ is finite, and is nonempty only if the rank of Σ_0 is equal to that of Y. This leads to an almost explicit product formula for the formal dimension (the Plancherel mass) of π (see [Op4, Corollary 3.32, Theorem 5.6], and [Op5, Theorem 4.10]).

The *abstract Plancherel theorem for* $\mathcal{H}$ (see [Dix]) gives the existence of a unique nonnegative measure $\nu_{\mathrm{Pl}}(\cdot, q)$ on $\widehat{\mathcal{H}}$ such that

$$\tau(f) = \int_{\pi \in \widehat{\mathcal{H}}} \mathrm{Tr}(\pi(f)) \, \mathrm{d}\nu_{\mathrm{Pl}}(\pi, q), \quad \text{for all } f \in \mathcal{H}. \tag{4.9.5}$$

The measure $\nu_{\mathrm{Pl}}(\cdot, q)$ on $\widehat{\mathcal{H}}$ is called the canonical *Plancherel measure* for $\mathcal{H}$.

Following in [HO, (1.21)], we consider the central idempotent T_0^+ in the Iwahori–Hecke algebra $\mathcal{H}(W_0, q)$ defined by

$$T_0^+ := \frac{1}{P_{W_0}(q)} \left(\sum_{w \in W_0} T_w \right). \tag{4.9.6}$$

Then T_0^+ corresponds to the representation triv of $\mathcal{H}(W_0, q)$

$$\mathrm{triv} \colon \begin{array}{ccc} \mathcal{H}(W_0, q) & \longrightarrow & \mathbb{C}^\times \\ T_w & \mapsto & q^{\ell(w)}. \end{array}$$

The subspace $T_0^+ C(\mathcal{H}) T_0^+$ is the space of all spherical functions in $C(\mathcal{H})$.

We set

$$\widehat{\mathcal{H}}_{\mathrm{sph}} := \left\{ \pi \in \widehat{\mathcal{H}} \, : \, [\pi_{\mathcal{H}(W_0,q)} : \mathrm{triv}] = 1 \right\}.$$

4.9.2 The formula in Theorem 4.1.2 and its generalisation

A different proof of the formula

The combination of Lemma 2.28 in [Op3] and Equations (1.23) and (2.9) in [HO] provides a different and very natural proof of (4.1.2), as explained below.

The inverse of the Satake isomorphism expresses that multiplication by T_0^+ defines an isomorphism of commutative algebras

$$\begin{array}{ccc} \mathbb{C}[\Lambda]^{W_0} & \xrightarrow{\sim} & T_0^+\mathcal{H}(W_0,q)T_0^+ \\ f & \mapsto & T_0^+ f. \end{array} \tag{4.9.7}$$

The spherical subalgebra $T_0^+\mathcal{H}(W_0,q)T_0^+$ of $\mathcal{H}(W_0,q)$ has the vector space basis $T_0^+\theta_\lambda T_0^+$ with $\lambda \in \Lambda^{++}$.

It follows from [Op3, Lemma 2.28] that

$$T_0^+\theta_\lambda T_0^+ = \frac{1}{P_{W_0}(q^{-1})} T_0^+ \left(\sum_{w\in W_0} w\left(c(\cdot,q)\theta_\lambda\right) \right), \tag{4.9.8}$$

for $\lambda \in \Lambda^{++}$ and with the c-function given by

$$c(\cdot,q) := \prod_{a\in\Sigma_0} \frac{1-q^{-1}\theta_{-a^\vee}}{1-\theta_{-a^\vee}}. \tag{4.9.9}$$

The formula in (4.9.8) is that of [Op3, (1.23)], and (4.1.2) is an immediate consequence of it, as shown in [HO, (2.9)].

A generalization of the formula

This combinatorial formula generalizes the formula for Hall–Littlewood symmetric functions. In [Lus1] Lusztig showed that it can be written in terms of the affine Hecke algebra and that the "q-weight multiplicities" or generalized Kostka–Foulkes polynomials coming from these spherical functions are Kazhdan–Lusztig polynomials for the affine Weyl group.

The interplay between affine Hecke algebras and Macdonald spherical functions has proven very fruitful. For instance, affine Hecke algebras turn out to be instrumental in obtaining explicit combinatorial formulas for the monomial expansion and for the structure constants of the Macdonald spherical functions. For instance, in [Ra], Ram associated to an affine Weyl group an *affine apartment* (a tessellation of an affine Euclidean space by alcoves on which the affine Weyl group acts), defined an *alcove walk algebra* with a basis indexed by galleries of alcoves (the alcove walks) in the apartment, starting from the standard alcove, and proved that the corresponding affine Hecke algebra is a quotient of this alcove walk algebra. In the reverse direction, properties of Macdonald's spherical functions are fundamental tools in the harmonic analysis of the affine Hecke algebra (see [Op4]) and for the explicit computation of the Kazhdan–Lusztig basis for the spherical Hecke algebra.

4.9.3 Casselman's proof

A smooth representation (π, V) of G is *admissible* if the space

$$V^J := \{v \in V : \pi(j)v = v \text{ for all } j \in J\}$$

of J-fixed vectors has finite dimension for every compact open subgroup J of G.

Casselman showed in [Cas] (see also [Car]) how results from the general theory of admissible representations of G may be applied to give a new proof of the explicit formula for zonal spherical functions. We denote by p_χ the projection from $C_c^\infty(G)$ to the space of the representation $I(\chi) := \mathrm{i}_{L,P}^G(\delta_P^{1/2}\chi)$:

$$p_\chi(f)(g) := \int_L \int_U \delta_P^{1/2}(l)\,\chi^{-1}(l)\,f(lug)\,\mathrm{d}l\,\mathrm{d}p. \qquad (4.9.10)$$

Let χ_1 denote the characteristic function of K. We set $\phi_{K,\chi} := p_\chi(\chi_1)$. The contragredient representation of $I(\chi)$ is isomorphic to $I(\chi^{-1})$.

The matrix coefficients of the spherical principal series describe precisely the spectrum of the corresponding C^*-algebra generated by the bi-invariant functions of compact support, often called a Hecke algebra. The spectrum of the commutative Banach $*$-algebra of bi-invariant L^1 functions is larger; when G is a semisimple Lie group with maximal compact subgroup K, additional characters come from matrix coefficients of the complementary series, obtained by analytic continuation of the spherical principal series.

In [Cas], Casselman reproved the above formulas using relations between $\mathcal{H}(G, I)$ with I an Iwahori subgroup of G and the algebras of intertwining operators for these principal series representations. This point of view was taken further by Casselman and Shalika in [CS], where they explicitly computed the spherical Whittaker functions for a p-adic reductive group. It is known from the work of Jacquet and Rallis [JR] that every irreducible admissible representation π of $\mathrm{GL}_n(k)$ possesses at most one (up to scaling by a constant factor) Shalika model. (Uniqueness for Archimedean fields is proven by Ash and Ginzburg in [AG].)

For every reduced crystallographic root system, van Diejen and Emsiz introduced in [DE] a unitary representation of the (extended) affine Hecke algebra given by discrete difference-reflection operators acting on a Hilbert space of complex functions on the weight lattice. They proved that the action of the center under this representation is diagonal on the basis of Macdonald spherical functions, and, as an application, they computed an explicit Pieri formula for these spherical functions. Theorem 4.1.2 is notably an ingredient of the proof of [DE, Proposition 5.9].

4.9.4 The c-function

The c-function appeared in Harish-Chandra's article [8] about the asymptotics of zonal spherical functions on Riemannian symmetric spaces of non-compact type and its applications to the Plancherel formula on these spaces. In [Hel1], Helgason described the importance of this function, notably in the study of spherical functions, eigenfunctions of invariant differential operators, Plancherel measures, Fourier transforms, orbital integrals, and symmetric spaces, see also [Gi]. For interesting applications of the c-function defined in this chapter, see [CR].

4.9.5 Some other results on spherical functions

For G a p-adic reductive and K a maximal compact subgroup of G, Tadić introduced in [Tad] a family of Schwartz spaces $C^{\gamma}(G,K)$, for $0<\gamma\leq 2$, of rapidly decreasing K-bi-invariant functions on G, which are Fréchet algebras under convolution, and used the results of this chapter to study spherical transformations on these spaces.

Finis and Matz studied in [FM] the asymptotic behavior of the traces of Hecke operators for spherical discrete automorphic representations of fixed level on general split reductive groups. Under a condition on the analytic behavior of intertwining operators, which is known in a large number of cases, they obtained the expected asymptotics in terms of the spherical Plancherel measure and an explicit estimate for the remainder.

Peterson studied in [Pet] eigenfunctions of the spherical Hecke algebra acting on $L^2(\Gamma_n\backslash G/K)$, where $G=\mathrm{PGL}_3(k)$ with k a p-adic field, $K=\mathrm{PGL}_3(\mathfrak{o}_k)$, and (Γ_n) is a sequence of cocompact torsion-free lattices, and proved a form of equidistribution on average for eigenfunctions whose spectral parameters lie in the tempered spectrum when the associated sequence of quotients of the Bruhat–Tits building Benjamini–Schramm converges to the building itself.

In [HSS], in the case of $G=\mathrm{PGL}_2(\mathbb{Q}_p)$, a quantum-probabilistic interpretation of the spherical Hecke algebra is given, as well as a new proof of the Fourier inversion formula.

The Bernstein centre opened the possibility of also studying generalized spherical functions in the p-adic case. Fix a minimal parabolic subgroup P_0 defined over F and let K be a maximal compact subgroup of G satisfying $G=P_0K$. Generalized spherical functions on G are eigenfunctions for the action of the Bernstein centre, which satisfy a transformation property for the action of K:

Definition Let (τ_1,W_1) and (τ_2,W_2) be finite-dimensional representations of K. A *generalized spherical function* of type (τ_1,τ_2) is a mapping $f\colon G\to\mathrm{Hom}_{\mathbb{C}}(W_2,W_1)$ which satisfies

$$f(k_1gk_2)=\tau_1(k_1)f(g)\tau_2(k_2),\quad \text{for all } k_1,k_2\in K,\ g\in G, \tag{4.9.11}$$

and which are eigenfunctions for the action of the Bernstein center.

In [HuTa], Huang and Tadić defined and studied basic properties of generalized spherical functions in the case of a reductive p-adic group (they are defined as eigenfunctions of the Bernstein center (see §2.9.5), which satisfy a transformation property for the action of a maximal compact subgroup). They proved that the spaces of generalized spherical functions are finite-dimensional, computed the dimensions of spaces of generalized spherical functions on a Zariski open dense set of infinitesimal characters, and, as a consequence, showed that on this Zariski open dense set, the dimension of the space of generalized spherical functions is constant on each connected component of infinitesimal characters. They also obtained the formula for the generalized spherical functions by integrals of Eisenstein type.

4.9.6 Buildings and Hecke algebras

In [Par1], Parkinson demonstrated a close connection between buildings and Hecke algebras through the combinatorial study of certain algebras of averaging operators associated to buildings. These results also have applications to the theory of random walks on buildings, as shown in [Par2] (see also [AST]). A set $\mathcal{C}$ is called a *chamber system* over a set I if each element i of I determines a partition of $\mathcal{C}$, two elements in the same block of this partition being said to be *i-adjacent*. The elements of $\mathcal{C}$ are called *chambers*, and we write $c \sim_i c'$ to mean that the chambers c and c' are i-adjacent. The building is said to be *thick* if for each $c \in \mathcal{C}$ and $i \in I$ there exist at least two distinct chambers $c' \neq c$ such that $c' \sim_i c$. The *rank* of a building is the cardinality of the index set I.

A building $\mathcal{B}$ may be viewed as a set $\mathcal{C}$ of chambers with an associated Coxeter system (W, S) and a W-distance function $\delta\colon \mathcal{C}\times\mathcal{C} \longrightarrow W$. For each $c \in \mathcal{C}$ and $w \in W$, we write

$$\mathcal{C}_w(c) := \{c' \in \mathcal{C} : \delta(c,c') = w\}. \tag{4.9.12}$$

An important assumption made in [Par1] is that the building is *regular*, that is, for every $s \in S$, we have

$$|\mathcal{C}_s(c)| = |\mathcal{C}_s(c)| \quad \text{for all } c, c' \in \mathcal{C}.$$

In a regular building we write $q_s := |\mathcal{C}_s(c)|$, and call the set $\{q_s\}_{s\in S}$ the *parameter system* of the building. Regularity implies the stronger result that $|\mathcal{C}_w(c)| = |C_w(c')|$ for all $c, c' \in \mathcal{C}$ and $w \in W$ (see [Par1, Proposition 2.1]), and as such we set $q_w := |\mathcal{C}_w(c)|$. All thick buildings with no rank 2 residues of type $\widetilde{\mathrm{A}}_1$ are regular (see [Par1, Theorem 2.4]).

Let $\mathcal{B}$ be an affine building of type Σ_0 with vertex set $\mathcal{V}$, and let $S := S(\Sigma_0)$. Let $\mathcal{A}$ be an apartment of $\mathcal{B}$. An isomorphism $\psi\colon \mathcal{A} \longrightarrow S$ is called *type rotating* if it is of the form $\psi = w \circ \psi_0$, where $\psi_0\colon \mathcal{A} \longrightarrow S$ is a *type preserving* isomorphism, and $w \in W^{\mathrm{e}}$.

The set Λ of coweights of Σ is a subset of $\mathcal{V}$, and we call elements of Λ the *good* vertices of $\mathcal{B}$. Let $\mathcal{V}_\Lambda$ denote the set of all good vertices of $\mathcal{B}$. Given $x \in \mathcal{V}_\Lambda$ and $\lambda \in \Lambda^{++}$, let $\mathcal{V}_\lambda(x)$ be the set of all $y \in \mathcal{V}_\Lambda$ such that there exists an apartment $\mathcal{A}$ containing x and y, and a type rotating isomorphism $\psi : \mathcal{A} \to S$ such that $\psi(x) = 0$ and $\psi(y) = \lambda$.

For $\lambda \in \Lambda^{++}$, let $\lambda^* := -w_0\lambda$. We have $\lambda^* \in \Lambda^{++}$, and that $y \in \mathcal{V}_\lambda(x)$ if and only if $x \in \mathcal{V}_{\lambda^*}(y)$ (see [Par1, Proposition 5.8]).

We have $|\mathcal{V}_\lambda(x)| = |\mathcal{V}_\lambda(x')|$ for all $x, x' \in \mathcal{V}_\Lambda$ and $\lambda \in \Lambda^{++}$ (see [Par1, Theorem 5.15]). We denote this common value by N_λ. We have

$$N_\lambda = \frac{P_{W_0}(q^{-1})}{P_{W_{0\lambda}}(q^{-1})} q(t_\lambda), \tag{4.9.13}$$

where $W_{0\lambda} := \{w \in W_0 : w\lambda = \lambda\}$, and $t_\lambda \in W^{\mathrm{e}}$ is the translation defined by $t_\lambda(v) = v + \lambda$, for all $v \in V$ (see [Par2, Proposition 1.5]).

For every $\lambda \in P+$, we define an operator A_λ, acting on the space of functions $f : \mathcal{V}_\Lambda \to \mathbb{C}$ by

$$(A_\lambda f)(x) := \frac{1}{N_\lambda} \sum_{y \in \mathcal{V}_\lambda(x)} f(y), \quad \text{for all } x \in \mathcal{V}_\Lambda, \tag{4.9.14}$$

and denote by $\mathfrak{A}$ the linear span of $\{A_\lambda\}_{\lambda \in \Lambda^{++}}$ over $\mathbb{C}$. For every $\lambda, \mu \in \Lambda^{++}$, we have

$$A_\lambda A_\mu = \sum_{\nu \in \Lambda^{++}} a_{\lambda,\mu;\nu} A_\nu, \tag{4.9.15}$$

where $a_{\lambda,\mu;\nu} \in \mathbb{Q}^+$ and $\sum_{\nu \in \Lambda^{++}} a_{\lambda,\mu,\nu} = 1$ (see [Par1, Corollary 5.22]), so that $\mathfrak{A}$ is an algebra. Moreover, we have

$$a_{\lambda,\mu;\nu} = \frac{N_\nu}{N_\lambda N_\mu} |\mathcal{V}_\lambda(x) \cap \mathcal{V}_{\mu^*}(y)||, \tag{4.9.16}$$

where $x, y \in \mathcal{V}_\Lambda$ are any vertices such that $y \in \mathcal{V}_\nu(x)$.

We set

$$\begin{aligned} \Sigma_{0,\mathrm{nm}} &:= \{a \in \Sigma_0 \ : \ 2a \notin \Sigma_0\}, \\ \Sigma_{0,\mathrm{nd}} &:= \left\{a \in \Sigma_0 \ : \ \frac{a}{2} \notin \Sigma_0\right\}, \quad \text{and} \\ \Sigma_{0,\mathrm{r}} &:= \Sigma_{0,\mathrm{nm}} \cap \Sigma_{0,\mathrm{nd}}. \end{aligned}$$

Hence we have $\Sigma_0 = \Sigma_{0,\mathrm{r}} \sqcup \Sigma_{0,\mathrm{nm}} \backslash \Sigma_{0,\mathrm{r}} \sqcup \Sigma_{0,\mathrm{nd}} \backslash \Sigma_{0,\mathrm{r}}$. We define

$$\mathbf{q}_a := \begin{cases} q(a) & \text{if } a \in \Sigma_{0,\mathrm{r}}, \\ q(0) & \text{if } a \in \Sigma_{0,\mathrm{nm}} \backslash \Sigma_{0,\mathrm{r}}, \\ q(a)q(0)^{-1} & \text{if } a \in \Sigma_{0,\mathrm{nd}} \backslash \Sigma_{0,\mathrm{r}}. \end{cases} \tag{4.9.17}$$

By convention, we write $\mathbf{q}_a = 1$ if $a \notin \Sigma_0$.

For $\lambda \in \Lambda^{++}$ and $s \in \operatorname{Hom}(\Lambda, \mathbb{C}^\times)$, as in §4.1, we set

$$\mathbf{c}(s) := \prod_{a \in \Sigma_0^+} \frac{1 - \mathbf{q}_a^{-1} \mathbf{q}_{a/2}^{-1/2} s(-a^\vee)}{1 - \mathbf{q}_{a/2}^{-1/2} s(-a^\vee)}. \tag{4.9.18}$$

Following [Par2, (2.1)], we define

$$P_\lambda(s) := \frac{q(t_\lambda)^{-1/2}}{P_{W_0}(q^{-1})} \sum_{w \in W_0} \mathbf{c}(ws)\, s(w\lambda), \tag{4.9.19}$$

provided that the denominators of the $c(ws)$ do not vanish. Since $P_\lambda(s)$ is a Laurent polynomial, these "singular" cases can be obtained from the generic formula by taking an appropriate limit as in §4.6.

Let $\omega_s \colon \mathfrak{A} \to \mathbb{C}$ be the linear map such that $\omega_s(A_\lambda) = P_\lambda(u)$ for every $\lambda \in \Lambda^{++}$. Theorem 3.3.12 admits the following generalization (see [Par2, Proposition 2.1]):

(1) Every algebra homomorphism $\omega \colon \mathfrak{A} \to \mathbb{C}$ is of the form $\omega = \omega_s$ for some $s \in \operatorname{Hom}(\Lambda, \mathbb{C}^\times)$,
(2) $\omega_s = \omega_{s'}$ if and only if $s' = ws$ for some $w \in W_0$.

Let $\mathbb{C}[\Lambda]$ denote the group algebra of Λ over $\mathbb{C}$, with the group operation written multiplicatively. Thus the elements of Λ are linear combination of the formal exponentials $\{\theta^\lambda\}_{\lambda \in \Lambda}$. The group W_0 acts on $\mathbb{C}[\Lambda]$ by linearly extending $w\theta^\lambda = \theta^{w\lambda}$, and we denote by $\mathbb{C}[\Lambda]^{W_0}$ the subalgebra consisting of all those $f \in \mathbb{C}[\Lambda]$ such that $wf = f$ for any $w \in W_0$. The map $A_\lambda \mapsto P_\lambda(\theta)$ determines an algebra isomorphism (see [Par2, Theorem 6.16]), and hence $\mathfrak{A} \cong \mathbb{C}[\Lambda]^{W_0}$.

4.9.7 An application of Theorem 4.7.1 to Eisenstein series

Let X be a bounded symmetric domain, let H be the group of holomorphic automorphisms of X, and let Γ be an arithmetic subgroup of H. Let F be a boundary component of X, rational (relative to Γ), with normalizer $P \subset H$. Theorem 2.5.6 in [Ha] uses Theorem 4.7.1 to show that Eisenstein series in the range of absolute convergence for P are completely characterized by their Hecke eigenvalues at good primes (see [Ha, Theorem 1.5]).

Chapter V
Plancherel measure

5.1 The standard case

Let $\widehat{\Lambda} = \mathrm{Hom}(\Lambda, \mathbb{R}/2\pi i\mathbb{Z})$ be the character group of the discrete group Λ. $\widehat{\Lambda}$ is the product of ℓ circles, and may be identified[1] with the torus V^*/L, where as in (3.3.13) L is the lattice of linear forms u on V such that $u(a^\vee) \in \mathbb{Z}$ for all $a \in \Sigma_0$. Let $\mathrm{d}s$ be Haar measure on the compact group $\widehat{\Lambda}$, normalized so that the total mass of Λ is 1.

We shall assume in this section that

$$q_{a/2} \geq 1, \forall a \in \Sigma_0. \tag{5.1.1}$$

Call this the *standard case*. The exceptional case, where $q_{a/2} < 1$ for some $a \in \Sigma_0$, will be examined separately in §5.2. Here we shall prove

Theorem 5.1.2 *Assume that $q_{a/2} \geq 1$ for all $a \in \Sigma_0$. Then the Plancherel measure μ (1.5.1) on the space Ω^+ of positive definite spherical functions on G relative to K is concentrated on the set $\{\omega_s : s \in \widehat{\Lambda}\}$ and is given there by*

$$\mathrm{d}\mu(\omega_s) = \frac{P_{W_0}(q^{-1})}{|W_0|} \cdot \frac{\mathrm{d}s}{|c(s)|^2}$$

where $|W_0|$ is the order of the Weyl group W_0.

Proof For $f_1, f_2 \in \mathcal{H}(G, K)$ we define

$$\langle f_1, f_2 \rangle = \int_G f_1(g)\overline{f_2(g)}\mathrm{d}g,$$

$$\langle \hat{f}_1, \hat{f}_2 \rangle = \frac{P_{W_0}(q^{-1})}{|W_0|} \int_{\widehat{\Lambda}} \hat{f}_1(\omega_s)\overline{\hat{f}_2(\omega_s)} \frac{\mathrm{d}s}{|c(s)|^2}.$$

[1] Therefore $V^*/L \cong \mathrm{Hom}(\Lambda, \mathbb{R}/2\pi i\mathbb{Z}) = \widehat{\Lambda} \cong S_u$.

I. G. Macdonald and A.-M. Aubert, *Spherical Functions on a Group of p-adic Type*, Lecture Notes in Mathematics 2392,
https://doi.org/10.1007/978-3-032-15671-6_5

We have to show that $\langle f_1, f_2\rangle = \langle \hat{f}_1, \hat{f}_2\rangle$. By linearity it is enough to take f_1, f_2 to be characteristic functions χ_{t_1}, χ_{t_2} $(t_1, t_2 \in \Lambda^{++})$, and we may assume that $t_1 \geq t_2$ with respect to the total ordering defined in (3.3.9). Clearly we have

$$\langle \chi_{t_1}, \chi_{t_2}\rangle = \begin{cases} 0 & \text{if } t_1 > t_2 \\ (Kt_1K : K) & \text{if } t_1 = t_2. \end{cases} \tag{5.1.3}$$

Now consider $\langle \hat{\chi}_{t_1}, \hat{\chi}_{t_2}\rangle$. To compute this we shall express $|c(s)|^{-2}\hat{\chi}_{t_1}(\omega_s)$ as an infinite ascending series, and $\hat{\chi}_{t_2}(\omega_s)$ as a descending polynomial, with at most one term in common. Term by term integration over the torus $\widehat{\Lambda}$ will then give the required result.

Since $\bar{s} = s^{-1}$ we have $\overline{c(b, s)} = c(b, s^{-1}) = c(-b, s)$ for each $b \in \Sigma_1$, hence

$$|c(s)|^{-2} = \left(c(s)c(s^{-1})\right)^{-1} = \prod_{b\in\Sigma_1} c(b, s)^{-1},$$

the product being over all the roots, positive and negative. This shows that $|c(s)|^{-2}$ is W_0-invariant, hence

$$|c(s)|^{-2} = \left(c(ws)c(ws^{-1})\right)^{-1} \tag{5.1.4}$$

for all $w \in W_0$.

Now

$$\hat{\chi}_t(\omega_s) = \int_G \chi_t(g)\omega_s(g^{-1})\mathrm{d}g = v_t\omega_s(z^{-1})$$

where $v_t = (KtK : K)$, and z is any element of $\nu^{-1}(t)$. Hence if s is nonsingular we have from (4.1.2) and (5.1.4)

$$\frac{1}{v_t}\cdot\frac{\hat{\chi}_t(\omega_s)}{|c(s)|^2} = \frac{\delta(t)^{-1/2}}{P_{W_0}(q^{-1})}\sum_{w\in W_0}\frac{(ws, t)}{c(ws^{-1})}$$

or equivalently

$$\frac{1}{v_t}\cdot\frac{\hat{\chi}_t(\omega_s)}{|c(s)|^2} = \frac{\delta(t)^{-1/2}}{P_{W_0}(q^{-1})}\sum_{w\in W_0}(ws, t)\prod_{b\in\Sigma_1^+}\frac{1 - q_{b/2}^{-1/2}(ws, t_b)}{1 - q_{b/2}^{-1/2}q_b^{-1}(ws, t_b)}. \tag{5.1.5}$$

Now if $b = a \in \Sigma_0$, then $q_{b/2}^{1/2}q_b = (q_aq_{a+1})^{1/2} > 1$, and if $b = a/2$ then $q_{b/2}^{1/2}q_b = q_{a/2} \neq 1$ (because $b \in \Sigma_1$). Hence the right-hand side of (5.1.5) is defined for *all* $s \in \widehat{\Lambda}$, and therefore (5.1.5) is valid for all $s \in \widehat{\Lambda}$.

Moreover, under the assumption (5.1.1) we have $q_{b/2}^{1/2} q_b > 1$ for all $b \in \Sigma_1$, and therefore

$$\prod_{b \in \Sigma_1^+} \frac{1 - q_{b/2}^{-1/2}(ws, t_b)}{1 - q_{b/2}^{-1/2} q_b^{-1}(ws, t_b)} = \prod_{b \in \Sigma_1^+} \left\{ 1 + (1 - q_b) \sum_{n=1}^{\infty} q_{b/2}^{-n/2} q_b^{-n} (ws, t_b)^n \right\},$$

these series being absolutely and uniformly convergent on the compact group $\widehat{\Lambda}$. Multiplying them all together we shall get from (5.1.5)

$$\frac{1}{v_t} \cdot \frac{\hat{\chi}_t(\omega_s)}{|c(s)|^2} = \frac{\delta(t)^{-1/2}}{P_{W_0}(q^{-1})} \sum_{t' \in \Lambda^{++}} a_{tt'} \sum_{w \in W_0} (ws, t) \tag{5.1.6}$$

say, with coefficients $a_{tt'}$ independent of s, and the summation being over all $t' \in \Lambda^{++}$ of the form

$$t' = t \cdot \prod_{b \in \Sigma_1^+} t_b^{n_b}$$

with integer exponents $n_b \geq 0$. It follows that $t' \geq t$ with respect to the ordering (3.3.9); also it is easy to see that $a_{tt} = 1$, so that (replacing t by t_1) we have

$$\frac{1}{v_{t_1}} \cdot \frac{\hat{\chi}_{t_1}(\omega_s)}{|c(s)|^2} = \frac{\delta(t_1)^{-1/2}}{P_{W_0}(q^{-1})} \left(\sum_{w \in W_0} (ws, t_1) + \sum_{\substack{t_1' > t_1 \\ t_1' \in \Lambda^{++}}} a_{t_1 t_1'} \sum_{w \in W_0} (ws, t_1') \right) \tag{5.1.6'}$$

for all $t_1 \in \Lambda^{++}$, the series on the right being uniformly convergent on $\widehat{\Lambda}$.

On the other hand, from (3.3.4′) and (3.3.8′),

$$\hat{\chi}_{t_2}(\omega_s) = s(\tilde{\chi}_{t_2}) = \delta(t_2)^{1/2}(s, < t_2 >) + \sum_{\substack{t_2' < t_2 \\ t_2' \in \Lambda^{++}}} \hat{\chi}_{t_2}(t_2')(s, < t_2' >). \tag{5.1.7}$$

Furthermore, we have

$$\int_{\widehat{\Lambda}} (ws, t_1') \overline{(s, < t_2' >)} \mathrm{d}s = \begin{cases} 0 & \text{if } t_1' \neq t_2', \\ 1 & \text{if } t_1' = t_2' \end{cases} \tag{5.1.8}$$

by orthogonality of characters on $\widehat{\Lambda}$. Hence, multiplying the infinite series (5.1.6′) by the polynomial (5.1.7) and integrating term by term over $\widehat{\Lambda}$, we shall have, using (5.1.8),

$$\frac{1}{v_{t_1}} \int_{\widehat{\Lambda}} \hat{\chi}_{t_1}(\omega_s) \overline{\hat{\chi}_{t_2}(\omega_s)} |c(s)|^{-2} \mathrm{d}s = \begin{cases} 0 & \text{if } t_1 > t_2, \\ \dfrac{|W_0|}{P_{W_0}(q^{-1})} & \text{if } t_1 = t_2. \end{cases}$$

Hence

$$\langle \hat{\chi}_{t_1}, \hat{\chi}_{t_2} \rangle = 0 = \langle \chi_{t_1}, \chi_{t_2} \rangle \text{ if } t_1 > t_2, \text{and}$$
$$\langle \hat{\chi}_{t_1}, \hat{\chi}_{t_1} \rangle = (Kt_1K : K) = \langle \chi_{t_1}, \chi_{t_1} \rangle. \qquad \square$$

5.2 The exceptional case

In this section we shall consider the case excluded from the considerations of §5.1, namely where $q_{a/2} < 1$ for some $a \in \Sigma_0$. First of all, this implies that $\frac{a}{2} \in \Sigma_1$ for some $a \in \Sigma_0$, so that the root system Σ_1 is not reduced. Since it is irreducible, it must be of type BC_ℓ and can therefore be described as follows. There exists an orthonormal basis $x_1, \ldots, x_\ell$ of A such that, if $e_1, \ldots, e_\ell \in V^*$ are the coordinate functions relative to this basis, the root system Σ_1 consists of

$$\pm e_i \ (1 \leq i \leq l), \quad \pm 2e_i \ (1 \leq i \leq l), \quad \pm e_i \pm e_j \ (1 \leq i < j \leq \ell)$$

and Σ_0 consists of

$$\pm 2e_i \ (1 \leq i \leq \ell), \quad \pm e_i \pm e_j \ (1 \leq i < j \leq l)$$

and is of type C_ℓ. We choose the set Π_0 of simple roots as follows:

$$\Pi_0 = \{e_1 - e_2, e_2 - e_3, \ldots, e_{\ell-1} - e_\ell, 2e_\ell\}.$$

The Weyl group W_0 is the group of all signed permutations of $e_1, \ldots, e_\ell$, i.e. $W_0 \cong \{\pm 1\}^\ell \rtimes S_\ell$, and has order $2^l l!$.

Let

$$q_0 = q_{\pm e_i \pm e_j}, \quad q_1 = q_{\pm e_i}, \quad q_2 = q_{\pm 2e_i},$$

so that $q_1 < 1$.

Next consider the translation group Λ. Put $t_i = t_{2e_i}$ $(1 \leq i \leq l)$. Then $t_i(0) = (2e_i)^\vee = x_i$, and hence $t_{\pm e_i \pm e_j} = t_i^{\pm 1} t_j^{\pm 1}$. The t_i form a basis of Λ.

If $s \in S$, put $s_i = s(t_i) \in \mathbb{C}^\times$. Then $c(s)$ is the product of the factors

$$\frac{(1 + q_1^{-1/2} s_i^{-1})(1 - q_1^{-1/2} q_2^{-1} s_i^{-1})}{1 - s_i^{-2}}$$

for $i = 1, \ldots, \ell$, and the factors

$$\frac{1 - q_0^{-1} s_i^{-1} s_j}{1 - s_i^{-1} s_j} \cdot \frac{1 - q_0^{-1} s_i^{-1} s_j^{-1}}{1 - s_i^{-1} s_j^{-1}}$$

for $1 \leq i < j \leq \ell$.

Finally, let $\phi(s) = c(s)c(s^{-1})$. Then we have to compute the integral

$$\frac{P_{W_0}(q^{-1})}{|W_0|} \int_{\widehat{\Lambda}} \hat{\chi}_{t_1}(\omega_s)\overline{\hat{\chi}_{t_2}(\omega_s)} \cdot \phi(s)^{-1} \mathrm{d}s \tag{5.2.1}$$

where $t_1, t_2 \in \Lambda^{++}$ and $t_1 \geq t_2$. Proceeding as in §5.1, we have

$$\hat{\chi}_{t_1}(\omega_s) = v_{t_1} \cdot \frac{\delta(t_1)^{-1/2}}{P_{W_0}(q^{-1})} \sum_{w \in W_0} c(ws)(ws, t_1).$$

Since the Haar measure $\mathrm{d}s$ on $\widehat{\Lambda}$ is invariant under W_0, it follows that the integral (5.2.1) is equal to

$$v_{t_1} \cdot \delta(t_1)^{-1/2} \int_{\widehat{\Lambda}} c(s)(s, t_1) \cdot \overline{\hat{\chi}_{t_2}(\omega_s)} \phi(s)^{-1} \mathrm{d}s.$$

On the other hand, from (3.3.8′),

$$\hat{\chi}_{t_2}(\omega_s) = \sum_{\substack{t_2' \leq t_2 \\ t_2' \in \Lambda^{++}}} \tilde{\chi}_{t_2}(t_2')(s, < t_2' >),$$

so that (5.2.1) now takes the form

$$v_{t_2} \cdot \delta(t_1)^{-1/2} \sum_{\substack{t_2' \leq t_2 \\ t_2' \in \Lambda^{++}}} \tilde{\chi}_{t_2}(t_2') \int_{\widehat{\Lambda}} c(s)(s, t_1)(s^{-1}, < t_2' >) \phi(s)^{-1} \mathrm{d}s. \tag{5.2.2}$$

In this expression, $(s^{-1}, < t_2' >)$ is a sum of terms $(s^{-1}, wt_2'w^{-1})$ $(w \in W_0)$. Since $t_1 \geq t_2 \geq t_2' \geq wt_2'w^{-1}$ for all $w \in W_0$, in order to compute (5.2.2) we have to compute the integrals of the form

$$I = \int_{\widehat{\Lambda}} c(s)(s, t)\phi(s)^{-1} \mathrm{d}s \tag{5.2.3}$$

where $t \in \Lambda$ and $t \geq 0$. Since $|s_i| = |s(t_i)| = 1$, we may express (5.2.3) as a multiple contour integral, each contour being the unit circle γ in the complex plane. If $\theta_i = \arg(s_i)$, we have

$$\mathrm{d}s = \prod_{j=1}^{l} (\mathrm{d}\theta_j / 2\pi)$$

and therefore

$$\mathrm{d}s = \prod_{j=1}^{l} \frac{1}{2\pi\sqrt{-1}} \frac{\mathrm{d}s_j}{s_j}.$$

Consequently, if $t = t_1^{\lambda_1} \dots t_\ell^{\lambda_\ell}$, the integral (5.2.3) takes the form

$$I = \frac{1}{2\pi\sqrt{-1}} \int_\gamma s_1^{\lambda_1 - 1} \mathrm{d}s_1 \dots \frac{1}{2\pi\sqrt{-1}} \int_\gamma s_\ell^{\lambda_\ell - 1} \mathrm{d}s_\ell \cdot \frac{1}{c(s^{-1})}. \tag{5.2.3$'$}$$

Since $t \geq 0$, the first of the exponents λ_i which is non-zero is positive.

We shall establish a reduction formula for I. For this purpose we have to introduce more notation. Let $J = (j_1, \dots, j_r)$ be a sequence of integrals satisfying

$$1 \leq j_1 < j_2 < \dots < j_r \leq l$$

and let $\widehat{\Lambda}_J$ be the set of all $s \in S$ such that

$$\begin{aligned} s_{j_\alpha} &= -q_0^{r-\alpha} q_1^{1/2}, \quad 1 \leq \alpha \leq r \\ |s_i| &= 1 \text{ if } i \notin J. \end{aligned}$$

Also define a function $\phi_J(s)$ inductively as follows:

$$\begin{aligned} \phi_\emptyset(s) &= \phi(s) = c(s)c(s^{-1}), \\ \phi_J(s) = \phi_{j_1,\dots,j_r}(s) &= \lim_{s_{j_1} \to -q_0^{r-1} q_1^{1/2}} \frac{\phi_{j_2,\dots,j_r}(s)}{1 + q_0^{1-r} q_1^{-1/2} s_{j_1}}, \quad r > 0. \end{aligned} \tag{5.2.4}$$

$\phi_J(s)$ is a function of the complex variables $s_i, i \notin J$, but we regard it as a function on $\widehat{\Lambda}_J$. It is easily checked from the product which defines $\phi(s)$ that $\phi_{j_2,\dots,j_r}(s)$ has $(1 + q_0^{1-r} q_1^{-1/2} s_{j_1})$ as a factor in the numerator, so that the right-hand side of (5.2.4) is well defined.

Let

$$I_J = I_{j_1,\dots,j_r} = \int_{\widehat{\Lambda}_J} c(s)(s,t)\phi_J(s)^{-1} \mathrm{d}s$$

where the measure $\mathrm{d}s$ on $\widehat{\Lambda}_J$ is given by

$$\mathrm{d}s = \prod_{j \notin J} \frac{1}{2\pi\sqrt{-1}} \frac{\mathrm{d}s_j}{s_j}.$$

($\widehat{\Lambda}_J$ is a translate of a subtorus of $\widehat{\Lambda}$, and this measure is the translation of the normalized Haar measure on that subtorus[2].)

Finally, let ϕ^1 (resp. c^1) denote the product obtained from ϕ (resp. c) by deleting all factors involving s_1, and let Λ^1 be the subgroup of Λ generated by $t_2, \dots, t_\ell$. Define $\widehat{\Lambda}_J^1$ and ϕ_J^1 as above for subsequences J of $\{2, 3, \dots, l\}$, and put

[2] From the definition we see that $\widehat{\Lambda}_J$ consists of the "zeros" of the c-functions, i.e. the "poles" of the μ-function. It is the (compact) subtorus in which the "real residual points" lie.

$$I_J^1 = \int_{\hat{\Lambda}_J^1} c^1(s)(s,t)\phi_J^1(s)^{-1}\mathrm{d}s$$

for $t \in \Lambda^1$, the measure here being $\mathrm{d}s = \prod_{j\neq 1} \frac{1}{2\pi\sqrt{-1}}\frac{\mathrm{d}s_j}{s_j}$.

The final form of the Plancherel measure will depend on how many of the numbers

$$q_1^{1/2}, q_0 q_1^{1/2}, \ldots, q_0^{l-1} q_1^{1/2}$$

are less than 1. So we define ϵ_r for $0 \le r \le \ell$ as follows: $\epsilon_0 = 1$ and

$$\epsilon_r = \begin{cases} 1 & \text{if } q_0^{r-1} q_1^{1/2} < 1 \\ 0 & \text{otherwise} \end{cases} \tag{5.2.5}$$

for $1 \le r \le \ell$.

Since $q_0 > 1$, it follows that $\epsilon_r = 1 \Rightarrow \epsilon_{r-1} = 1$.

Having set all this up, the key to the calculation of the integral I in (5.2.3) is the following lemma:

Lemma 5.2.6 *Let $J = (j_1, \ldots, j_r)$ be a subsequence of $(2, 3, \ldots, l)$ and put $J' = (1, j_1, \ldots, j_r)$. Then*

$$\epsilon_r I_J + \epsilon_{r+1} I_{J'} = \begin{cases} 1 & \textit{if } \lambda_1 > 0, \\ \epsilon_r I_J^1 & \textit{if } \lambda_1 = 0. \end{cases}$$

Proof If $\epsilon_r = 0$ then also $\epsilon_{r+1} = 0$ (because $q_0 > 1$) and there is nothing to prove. So assume that $\epsilon_r = 1$. Consider the integral I_J, and integrate it round the unit circle γ with respect to the variable s_1. To do this we must pick out the factors in the product $c(s)\phi_J(s)^{-1}$ which involves s_1. An elementary calculation shows that they are, after some cancellations,

$$\frac{1 - s_1^2}{(1 + q_0^{-r} q_1^{-1/2} s_1)(1 - q_1^{-1/2} q_2^{-1} s_1)} \cdot \frac{1 + q_0^{r-1} q_1^{1/2} s_1}{1 + q_0^{-1} q_1^{1/2} s_1} \cdot \prod_{j \notin J'} \frac{1 - s_1 s_j^{\pm 1}}{1 - q_0^{-1} s_1 s_j^{\pm 1}}. \tag{5.2.7}$$

Hence, since $|s_j| = 1$ for $j \notin J'$, the integrand

$$\frac{1}{2\pi\sqrt{-1}} \frac{c(s)(s,t)}{\phi_J(s)} \frac{\mathrm{d}s_1}{s_1}$$

considered as a function of s_1, has *poles* at the points

$$s_1 = -q_0^r q_1^{1/2},\ q_1^{1/2} q_2,\ -q_0 q_1^{-1/2},\ q_0 s_i^{\pm 1}$$

and also at the origin, if $\lambda_1 = 0$.

The points $q_1^{1/2}q_2$, $-q_0q_1^{-1/2}$, $q_0s_i^{\pm 1}$ all lie outside the unit circle γ (for $q_1^{1/2}q_2 = \delta^{1/2}(t_\ell) > 1$, and q_0 and $q_1^{-1/2}$ are both > 1). The point $-q_0^r q_1^{1/2}$ lies inside γ if and only if $\epsilon_{r+1} = 1$. (If $q_0^r q_1^{1/2} = 1$, then the factor $(1 + q_0^{-r} q_1^{-1/2} s_1)$ in the denominator of (5.2.7) cancels into the factor $1 - s_1^2$ in the numerator, so there is never a pole *on* the unit circle.) The lemma now follows by computing the residues at the poles in the usual way. □

Corollary 5.2.8

$$\sum_J \epsilon_{|J|} I_J = \begin{cases} 0 & \text{if } t > 1, \\ 1 & \text{if } t = 1, \end{cases}$$

where the sum is over all subsequences J of $(1, 2, \ldots, l)$ and $|J| = \mathrm{Card}(J)$.

Proof By induction on l. The induction can start with $l = 1$, when the result is easily verified.

If $\lambda_1 > 0$, we have from (5.2.6)

$$\sum_J \epsilon_{|J|} I_J = 0.$$

If $\lambda_1 = 0$, again from (5.2.6) we have

$$\sum_J \epsilon_{|J|} I_J = \sum_{J^*} \epsilon_{|J^*|} I^1_{J^*},$$

the sum on the right being over all subsequences J^* of $(2, 3, \ldots, l)$. The result now follows from the inductive hypothesis. □

Now let $\pi \colon S \to \Omega$ be the mapping $s \mapsto \omega_s$ of S onto the set Ω of all z.s.f. on G relative to K. The fibres of π are the orbits of the action of W_0 on S.[3] Clearly $\pi(\widehat{\Lambda}_r)$ depends only on the number of elements in J, and not on the particular sequence. Let

$$\Omega_r = \pi(\widehat{\Lambda}_J) \text{ if } |J| = r.$$

From the definition of $\widehat{\Lambda}_J$, the z.s.f. belonging to Ω_r are the ω_s for which r of the numbers $s(t_i)$ $(1 \le i \le l)$ take the values

$$-q_1^{1/2}, -q_0q_1^{1/2}, \ldots, -q_0^{r-1}q_1^{1/2}$$

(or their inverses), and the remaining $s(t_i)$ have absolute value 1. Since $\phi(s)$ is invariant under the action of W_0, it follows that $\phi_J(s)$ depends only on the number of elements in J; so we may write

$$\phi_r(\omega_s) = \phi_J(s) \text{ if } |J| = r.$$

[3] Recall that $\omega_s = \omega_{s'}$ if and only if $s' = ws$ for some $w \in W_0$. The (categorial) quotient $W_0\backslash S$ is in bijective correspondence to the semisimple conjugacy classes of the dual group.

Proposition 5.2.9 *The spherical functions belonging to* Ω_r *are positive definite if* $\epsilon_r = 1$.

We leave the proof until later.

We have already defined a measure $\mathrm{d}s$ on each $\widehat{\Lambda}_J$ giving it total mass 1. Now define a measure $\mathrm{d}\omega$ on Ω_r as follows:

$$\int_{\Omega_r} f(\omega)\mathrm{d}\omega = \int_{\pi^{-1}(\Omega_r)} f(\omega_s)\mathrm{d}s$$

for any continuous function f on Ω_r with compact support.

Theorem 5.2.10 *The Plancherel measure* μ *on the space of positive definite spherical functions is concentrated on the sets* Ω_r *such that* $\epsilon_r = 1$, *where* ϵ_r *is defined in* (5.2.5). *On* Ω_r, μ *is given by*

$$\mathrm{d}\mu(\omega) = \frac{P_{W_0}(q^{-1})}{w_r}\frac{\mathrm{d}\omega}{\phi_r(\omega)}$$

where $w_r = 2^{l-r}(l-r)!$ *is the order of the normalizer in* W_0 *of any of the sets* $\widehat{\Lambda}_J$ *such that* $|J| = r$.

Proof For $f_1, f_2 \in \mathcal{H}(G, K)$, define

$$\langle \hat{f}_1, \hat{f}_2\rangle_r = \int_{\Omega_r} \hat{f}_1(\omega)\overline{\hat{f}_2(\omega)}\mathrm{d}\mu(\omega) \quad (0 \le r \le l)$$

where $\mathrm{d}\mu(\omega)$ is as above. Then we have to show that

$$\sum_{r=0}^{l} \epsilon_r\langle \hat{f}_1, \hat{f}_2\rangle_r = \langle f_1, f_2\rangle.$$

As before, we take $f_1 = \chi_{t_1}, f_2 = \chi_{t_2}$ with $t_1, t_2 \in \Lambda^{++}$ and $t_1 \ge t_2$. Let J_r be the sequence $(1, 2, \ldots, r)$. Then exactly as in §5.1

$$\langle \hat{\chi}_{t_1}, \hat{\chi}_{t_2}\rangle_r = \frac{P_{W_0}(q^{-1})}{w_r}\int_{\widehat{\Lambda}_{J_r}} \hat{\chi}_1(\omega_s)\overline{\hat{\chi}_2(\omega_s)}\frac{\mathrm{d}s}{\phi_{J_r}(s)}$$

$$= \frac{v_{t_1}\cdot\delta(t_1)^{-1/2}}{w_r}\sum_{\substack{t_2'\le t_2\\ t_2'\in\Lambda^{++}}} \tilde{\chi}_{t_2}(t_2')\sum_{w\in W_0}\int_{\widehat{\Lambda}_{J_r}} c(ws)(ws, t_1)(s^{-1}, < t_2' >)\frac{\mathrm{d}s}{\phi_{J_r}(s)}.$$

Now it is easily verified that if $s \in \widehat{\Lambda}_{J_r}$ and $w \in W_0$, then $c(ws) = 0$ unless w maps $t_1, \ldots, t_r$ to $t_{j_1}, \ldots, t_{j_r}$ respectively with $j_1 < \cdots < j_r$. Hence

$$\begin{aligned}\frac{1}{w_r} \sum_{w \in W_0} \int_{\widehat{\Lambda}_{J_r}} c(ws)(ws, t_1)(s^{-1}, < t_2' >) \frac{\mathrm{d}s}{\phi_{J_r}(s)} \\ = \sum_{|J|=r} \int_{\widehat{\Lambda}_J} c(s)(s, t_1)(s^{-1}, < t_2' >) \frac{\mathrm{d}s}{\phi_J(s)}.\end{aligned}$$

Consequently

$$\begin{aligned}&\sum_{r=0}^{l} \epsilon_r \langle \hat{\chi}_{t_1}, \hat{\chi}_{t_2} \rangle_r \\ &= v_{t_1} \cdot \delta(t_1)^{-1/2} \sum_{\substack{t_2' \le t_2 \\ t_2' \in \Lambda^{++}}} \tilde{\chi}_{t_2}(t_2') \sum_J \epsilon_{|J|} \int_{\widehat{\Lambda}_J} \frac{c(s)(s, t_1)(s^{-1}, < t_2' >) \mathrm{d}s}{\phi_J(s)}\end{aligned}$$

and by (5.2.8) the inner sum is 0 or 1 according as $t_1 \neq t_2'$ or $t_1 = t_2'$. Hence, finally,

$$\sum_{r=0}^{l} \epsilon_r \langle \hat{\chi}_{t_1}, \hat{\chi}_{t_2} \rangle_r = \begin{cases} 0 & \text{if } t_1 > t_2 \\ \langle \chi_{t_1}, \chi_{t_1} \rangle & \text{if } t_1 = t_2 \end{cases}$$

since $v_{t_1} \cdot \delta(t_1)^{-1/2} \tilde{\chi}_{t_1}(t_1) = v_{t_1} = \langle \chi_{t_1}, \chi_{t_1} \rangle$. □

We have still to prove (5.2.9). Consider first the set Ω_ℓ: this consists of a single spherical function, say ω_{s_0}, where $s_0 \in S$ is given by

$$s_0(t_i) = -q_0^{i-1} q_1^{1/2} \quad (1 \le i \le l).$$

Lemma 5.2.11 *Let $z \in Z^{++}$, $v(z) = t$. Then $\omega_{s_0} = (\delta^{-1/2} s_0, t)$.*

Proof Consider $c(ws_0)$ where $w \in W_0$ and $w \neq 1$. One sees immediately that some factor in the numerator vanishes, so that $c(ws_0) = 0$ if $w \neq 1$. Also a simple verification shows that $c(s_0) = c(\delta^{1/2})$, and hence $c(s_0) = P_{W_0}(q^{-1})$ by (4.5.8). The lemma now follows from (4.1.2). □

Lemma 5.2.12 *If $\epsilon_\ell = 1$ then ω_{s_0} is square-integrable.*

Proof We have

$$\int_G |\omega_{s_0}(g)|^2 \mathrm{d}g = \sum_{t \in \Lambda^{++}} v_t (\delta^{-1/2} s_0, t)^2$$

where $v_t = (KtK : K) = \delta(t) Q^t(q^{-1})$ by (3.2.15). Hence we have to show that the series

$$\sum_{t\in\Lambda^{++}} Q^t(q^{-1})(s_0,t)^2 \tag{5.2.13}$$

converges if $|s_0(t_i)| < 1$ for $1 \le i \le l$.

Since $t \in \Lambda^{++}$ we have $t = \prod_{i=1}^{l} t_i^{\lambda_i}$ with $\lambda_1 \ge \lambda_2 \ge \cdots \ge \lambda_\ell \ge 0$. Let $\rho = (\rho_1, \ldots, \rho_r)$ be a sequence of integers such that

$$1 \le \rho_1 < \rho_2 < \cdots < \rho_r \le l \tag{5.2.14}$$

and consider the $t = \prod t_i^{\lambda_i}$ such that

$$\lambda_1 = \lambda_2 = \cdots = \lambda_{\rho_1} > \lambda_{\rho_1+1} = \cdots = \lambda_{\rho_2} > \cdots \lambda_{\rho_r+1} = \cdots = \lambda_\ell = 0.$$

$Q^t(q^{-1})$ will be the same for all these t, because they all have the same stabilizer W_{0t} in the Weyl group W_0. Hence the series (5.2.13) splits up into a sum of series, one for each sequence ρ satisfying (5.2.14), and each of these series is a product of geometric series whose common ratios are products of the $s_0(t_i)^2$. Hence all these series converge and therefore so does the series (5.2.13). □

Remark It is not difficult to show, using (5.2.11), that ω_{s_0} is a z.s.f. on G relative to the Iwahori subgroup B. By (1.2.6) the z.s.f. on G relative to B correspond bijectively to the $\mathbb{C}$-algebra homomorphisms $\mathcal{H}(G,B) \to \mathbb{C}$. Let Π_0 (resp. Π) be the set of simple roots for Σ_0 (resp. Σ), (so that $\Pi_0 = \{a_1, \ldots, a_\ell\}, \Pi = \{\alpha_0, \ldots, \alpha_\ell\}$, where $\alpha_0 = 1 - 2e_1, \alpha_i = a_i = e_i - e_{i+1}$ $(1 \le i \le l-1), \alpha_\ell = a_\ell = 2e_\ell)$. Then $\mathcal{H}(G,B)$ is generated as a $\mathbb{C}$-algebra by the characteristic functions χ_i of $Bw_{\alpha_i}B$ $(0 \le i \le l)$, and the homomorphism $\hat{\omega}_{s_0} : \mathcal{H}(G,B) \to \mathbb{C}$ is given by $\chi \mapsto q(w_{\alpha_i})$, $(1 \le i \le l)$, $\chi_0 \mapsto -1$.

In general, the z.s.f. on G relative to B correspond naturally to the linear characters of the Weyl group W. So they are always finite in number, equal to the order of W made abelian.

We shall now sketch a proof of (5.2.9). Consider a zonal spherical function $\omega_s \in \Omega_r$, where $\epsilon_r = 1$. We may assume that

$$s(t_i) = -q_0^{r-i}q_1^{-1/2} \quad (1 \le i \le r),$$
$$|s(t_i)| = 1 \text{ if } i > r.$$

Let A' (resp. A'') be the subspace of A spanned by the first r basis vectors $x_1, \ldots, x_r$ (resp. by $x_{r+1}, \ldots, x_r$). The restrictions to A' of the roots $a \in \Sigma_0$ (resp. affine roots $\alpha \in \Sigma$) which vanish identically on A'' form a subsystem Σ_0' of Σ_0 (resp. a subsystem Σ' of Σ). The Weyl group W_0' and affine Weyl group W' of Σ' can be regarded as subgroups of W_0 and W respectively. The translation subgroup Λ' of W' is generated by $t_1, \ldots, t_r$. Likewise, starting with A'' we define Σ_0'', Σ'', etc. (Note that $\Sigma \supsetneq \Sigma' \cup \Sigma''$ in general.)

Let G' be the subgroup of G generated by the U_α with $\alpha \in \Sigma$ and by $N' = \nu^{-1}(W')$, and define G'' similarly. Then $G'G''$ is a subgroup of G.

Let $s' = s|\Lambda'$, $s'' = s|\Lambda''$. Then $\omega_{s'}$ is a square-integrable spherical function on G' by (5.2.12), and hence by (1.4.8) is positive definite. Again, since s'' is a character of Λ'', it follows from (3.3.1) (applied to G'') that $\omega_{s''}$ is positive definite.

Now let P_0 be the (parabolic) subgroup of G generated by G', G'' and U^-, and let Δ_0 be the modulus function on P_0. P_0 is the semi-direct product of $G'G''$ and the subgroup U_0^- generated by the root subgroups $U_{(a)}$ corresponding to *negative* roots $a \in \Sigma_0$ which do not belong to either Σ_0' or Σ_0'' (i.e. $a = -ei \pm e_j$ with $i \leq r$ and $j > r$). The spherical function $\Delta_0^{-1/2}\omega_{s'}\omega_{s''}$ on $G'G''$ extends to a spherical function on P_0, say ω_0. By (1.4.7), the spherical function induced by ω_0 on G is positive definite, and by transitivity of induction (1.3.3) it is equal to ω_s. This completes the proof of (5.2.9) and hence of (5.2.10).

5.3 Comparison with the real and complex cases

Let k be a local field, that is to say a non-discrete locally compact field. Then k is either $\mathbb{R}$ or $\mathbb{C}$ or a non-Archimedean local field, and the additive group of k is self-dual. Associated canonically with k there is a meromorphic function $\gamma_k(s)$ of a complex variable s, sometimes called the gamma-function of k. We recall its definition briefly (see Tate's thesis [17], where it is denoted by $\rho(|\ |^s)$.)

If f is any well-behaved function on k, let $\hat{f}$ be its Fourier transform with respect to the additive group structure. Since k^+ is self-dual, $\hat{f}$ is a function on k^+. (The Haar measure is normalized so that $\hat{\hat{f}} = f(-x)$.) If $\mathrm{Re}(s) > 0$ define

$$\zeta(f, s) = \int_{k^\times} f(x)\|x\|^s \mathrm{d}^\times x,$$

where $\|x\|$ is the normalized absolute value on k (i.e. $\mathrm{d}(ax) = \|a\|\mathrm{d}x$ where $\mathrm{d}x$ is additive Haar measure) and $\mathrm{d}^\times x$ is a Haar measure on the multiplicative group $k^\times$. Then $\zeta(f, s)$ has a functional equation

$$\zeta(f, s) = \gamma_k(s)\zeta(\hat{f}, 1-s) \tag{5.3.1}$$

with $\gamma_k(s)$ independent of f. This equation defines $\gamma_k(s)$ for $\mathrm{Re}(s) \in (0, 1)$, and γ_k is then extended by analytic continuation to a meromorphic function in the whole complex plane.

We can compute $\gamma_k(s)$ from (5.3.1) by choosing the function f intelligently.

- $k = \mathbb{R}$: take $f(x) = e^{-\pi x^2}$, then $\hat{f} = f$ and one finds that

$$\gamma_{\mathbb{R}}(s) = \frac{\pi^{-s/2}\Gamma(s/2)}{\pi^{(s-1)/2}\Gamma((1-s)/2)}$$

so that

$$\gamma_{\mathbb{R}}(s)\gamma_{\mathbb{R}}(-s) = \mathrm{B}(\frac{1}{2}, \frac{s}{2})\mathrm{B}(\frac{1}{2}, -\frac{s}{2}) \tag{5.3.2$_{\mathbb{R}}$}$$

where B is Euler's beta-function.

• $k = \mathbb{C}$: take $f(x) = e^{-2\pi|x|^2}$ ($|x|$ the ordinary absolute value on $\mathbb{C}$: in fact $\|x\| = |x|^2$), then again $\hat{f} = f$ and we have

$$\gamma_{\mathbb{C}}(s) = \frac{(2\pi)^{-s}\Gamma(s)}{(2\pi)^{s-1}\Gamma(1-s)}$$

so that

$$\gamma_{\mathbb{C}}(s)\gamma_{\mathbb{C}}(-s) = -4\pi^2/s^2. \tag{5.3.2$_{\mathbb{C}}$}$$

• $k = p$-adic field. Taking f to be the characteristic function of the ring of integers of k, one finds that

$$\gamma_k(s) = d^{1-s}\frac{1-q^{-1-s}}{1-q^{-s}}$$

where d is the discriminant of k and q is the number of elements in the residue field. Hence in this case we have

$$\gamma_k(s)\gamma_k(-s) = d^{-1}\frac{1-q^{-1-s}}{1-q^{-s}} \cdot \frac{1-q^{-1+s}}{1-q^{s}}. \tag{5.3.2$_p$}$$

Now let G be the universal Chevalley group $G(\Sigma_0, k)$ as in §2.5, where k is any local field. If $k = \mathbb{R}$ or $\mathbb{C}$, let K be any maximal compact subgroup of G (they are all conjugate). If k is p-adic, let K be the maximal compact subgroup of G defined as hitherto in terms of the affine root structure on G given in §2.5.

If $k = \mathbb{R}$ or $\mathbb{C}$, the z.s.f. on G relative to K are parametrized by the $\mathbb{R}$-linear mappings $s: A \to \mathbb{C}$, and the Plancherel measure for the positive definite spherical functions is supported on the space of pure imaginary s, which is naturally a real vector space of dimension $\ell = \dim A$. If ds is a Euclidean measure on this space, then the Plancherel measure μ is of the form

$$\mathrm{d}\mu(\omega_s) = \frac{\kappa}{|c(s)|^2}\mathrm{d}s$$

where κ is a constant (depending on the choices of ds and of Haar measure on G) and

$$c(s) = \begin{cases} \prod_{a\in\Sigma_0^+} \mathrm{B}\left(\frac{1}{2}, \frac{1}{2}s(a^\vee)\right), & \text{if } k = \mathbb{R}, \\ \prod_{a\in\Sigma_0^+} s(a^\vee)^{-1}, & \text{if } k = \mathbb{C}. \end{cases}$$

Hence, from (5.3.2), we have

$$c(s)c(-s) = \prod_{a\in\Sigma_0} \gamma_k\left(s(a^\vee)\right) \tag{5.3.3$_{\mathbb{R},\mathbb{C}}$}$$

in both cases, apart from a constant factor in the case $k = \mathbb{C}$.

On the other hand, if k is p-adic, then as we have seen in §5.1 the support of the Plancherel measure is the character group $\widehat{\Lambda}$ of Λ. To bring out the analogy with the real and complex cases we shall replace the multiplicative parametrization of the spherical functions, which we have used hitherto, by an additive parametrization. If $s_0 \in S = \mathrm{Hom}(\Lambda, \mathbb{C}^*)$, we define $s \colon A \to \mathbb{C}/(\frac{2\pi i}{\log q})\mathbb{Z}$ by the rule[4]

$$s_0(t_a)^{-1} = q^{-s(a^\vee)}, \quad a \in \Sigma_0$$

where as before q is the number of elements in the residue field of k. Then from (5.1.2) the Plancherel measure μ is of the form

$$\mathrm{d}\mu(\omega_s) = \frac{\kappa}{|c(s)|^2}\mathrm{d}s$$

where $\mathrm{d}s$ is Euclidean measure on the space of pure imaginary s, κ is a constant, and

$$c(s) = \prod_{a \in \Sigma_0^+} \frac{1 - q^{-1-s(a^\vee)}}{1 - q^{-s(a^\vee)}}$$

(because in the present situation all the q_a are equal to q). Hence, from $(5.3.2_p)$ we have

$$c(s)c(-s) = \prod_{a \in \Sigma_0} \gamma_k\left(s(a^\vee)\right), \qquad (5.3.3_p)$$

apart from a constant factor depending only on k.

So, to sum up:

Theorem 5.3.4 *If k is any local field and $G = G(\Sigma_0, k)$ is a universal Chevalley group, then the Plancherel measure on the space of positive definite spherical functions on G relative to the maximal compact subgroup K is of the form*

$$\mathrm{d}\mu(\omega_s) = \frac{\kappa \cdot \mathrm{d}s}{\prod_{a \in \Sigma_0} \gamma_k\left(s(a^\vee)\right)}$$

where κ is a constant.

Also for non-split groups there is a strong resemblance between the Plancherel measure in the real case and in the p-adic case.

Let $\mathbf{G}$ be a simply-connected simple algebraic group defined over a local field k, and let $G = \mathbf{G}(k)$. (We can exclude the case $k = \mathbb{C}$ from here on, because in that case $\mathbf{G}$ is necessarily split.)

• $k = \mathbb{R}$. Let $G = KAN$ be an Iwasawa decomposition, $\mathfrak{g}$ and $\mathfrak{a}$ the Lie algebras of G and A respectively, Σ_1 the set of "restricted roots" of $\mathfrak{g}$ relative to $\mathfrak{a}$, and for each $b \in \Sigma_1$ let m_b be the multiplicity of b. Then the z.s.f. on G relative to K are parametrized by the elements λ of $\mathrm{Hom}_{\mathbb{R}}(\mathfrak{a}, \mathbb{C})$, and the Plancherel measure is

[4] Macdonald wrote $s_0(t_a)$ instead of its inverse.

supported on the subspace consisting of the pure imaginary $\lambda\colon \mathfrak{a} \to i\mathbb{R}$. Write ω_λ for the z.s.f. corresponding to λ. Then [9] the Plancherel measure μ is given by

$$\mathrm{d}\mu(\omega_\lambda) = \frac{\kappa \cdot \mathrm{d}\lambda}{|c(\lambda)|^2}$$

where κ is a constant and (see [5]) $c(\lambda)$ is a product of beta-functions, namely

$$c(\lambda) = \prod_{b \in \Sigma_1^+} \mathrm{B}\left(\frac{m_b}{2}, \frac{m_b}{4} + \frac{1}{2}\lambda(b^\vee)\right).$$

Let

$$\zeta_{\mathbb{R}}(s) = \pi^{-s/2}\Gamma\left(\frac{s}{2}\right), \quad s \in \mathbb{C} \tag{5.3.5}$$

which is the local zeta-function $\zeta(f, s)$ for the function $f(x) = e^{-\pi x^2}$. In this notation we have

$$c(\lambda) = \kappa \prod_{b \in \Sigma_1^+} \frac{\zeta_{\mathbb{R}}\left(\frac{1}{2}m_{b/2} + \lambda(b^\vee)\right)}{\zeta_{\mathbb{R}}\left(m_b + \frac{1}{2}m_{b/2} + \lambda(b^\vee)\right)} \tag{5.3.6}$$

where κ is independent of λ.

• k p-adic. If the number of elements in the residue field of k is q, then each index q_b $(b \in \Sigma_1)$ is a power of q. We shall therefore write $q_b = q^{m_b}$ $(b \in \Sigma_1)$ and call m_b the *formal multiplicity* of the root $b \in \Sigma_1$. Next, to bring our notation into line with that used above in the case $k = \mathbb{R}$, we shall replace the parameter $s \in \mathrm{Hom}(\Lambda, \mathbb{C}^*)$ for the spherical functions by $\Lambda \in \mathrm{Hom}_{\mathbb{R}}(A, \mathbb{C})$ defined by $s(t_a) = q^{\lambda(a^\vee)}$ for all $a \in \Sigma_0$. λ is not uniquely determined by s, but it is unique modulo the lattice $\frac{2\pi i}{\log q}L$, where as in §2.5 L is the lattice of all $u \in V^*$ such that $u(a^\vee) \in \mathbb{Z}$ for all $a \in \Sigma_0$. In particular, if $s \in \widehat{\Lambda}$ then λ is pure imaginary. Writing $c(\lambda)$ in place of $c(s)$, we have from §4.1

$$c(\lambda) = \prod_{b \in \Sigma_1^+} \frac{1 - q^{-\left(\frac{m_{b/2}}{2} + m_b + \lambda(b^\vee)\right)}}{1 - q^{-\left(\frac{m_{b/2}}{2} + \lambda(b^\vee)\right)}}.$$

Now let

$$\zeta_k(s) = (1 - q^{-s})^{-1}, \quad s \in \mathbb{C} \tag{5.3.7}$$

which is the local zeta-function $\zeta(f, s)$ for the characteristic function of the ring of integers of k. Then in this notation we have

Theorem 5.3.8 *For k real or p-adic, the Plancherel measure is*

$$\mathrm{d}\mu(\omega_\lambda) = \frac{\kappa \cdot \mathrm{d}\lambda}{|c(\lambda)|^2}$$

where κ is a constant and

$$c(\lambda) = \prod_{b \in \Sigma_1^+} \frac{\zeta_k\left(\frac{1}{2}m_{b/2} + \lambda(b^\vee)\right)}{\zeta_k\left(\frac{1}{2}m_{b/2} + m_b + \lambda(b^\vee)\right)}$$

the functions ζ_k being defined by (5.3.5) *and* (5.3.7).

Remark There is one important difference between the real and p-adic cases: in the p-adic case the formal multiplicity m_b can be *negative* if $2b$ is also a root. This is precisely the "exceptional case" dealt with in §5.2.

5.4 Notes on Chapter V

The fact that the support of the Plancherel measure can be bigger than $\widehat{\Lambda}$ was first remarked by Matsumoto [13]. I had neglected this possibility in my original calculations.

5.5 Additions to Chapter V

5.5.1 An application of Theorem 5.1.2

Let V be a finite-dimensional real vector space with a positive definite inner product $(\ ,\ )$, let Σ_0 be a root system in V, and let Λ be the coweight lattice as defined in (2.9.3). We fix a set $\Sigma_0^+ = \{a_1, \ldots, a_\ell\}$ of positive roots. Let q be an indeterminate. Following [Lus1], we define a q-analogue of Kostant's partition function as

$$\widehat{\mathcal{P}}(\lambda; q) := \sum_{\substack{(n_1,\ldots,n_\ell)\in\mathbb{Z}_{\geq 0}^m \\ n_1a_1+\cdots+n_\ell a_\ell=\lambda}} q^{n_1+\cdots+n_\ell}, \quad \text{for } \lambda \in \Lambda. \tag{5.5.1}$$

For $\lambda, \mu \in \Lambda$, we put

$$K_{\lambda,\mu}(q) := \sum_{w\in W} \mathrm{sgn}(w)\, \widehat{\mathcal{P}}(w(\lambda+\rho) - (\mu+\rho); q), \tag{5.5.2}$$

where $2\rho := \sum_{i=1}^{\ell} a_i$ and $\mathrm{sgn}\colon W \longrightarrow \{\pm 1\}$ is the sign character of W.

Let $\mathfrak{g}$ be a complex semisimple Lie algebra of type Σ_0, and let χ_λ an irreducible character of $\mathfrak{g}$ with highest weight λ ($\lambda \in \Lambda^{++}$). By using Theorem 5.1.2, Kato obtained in [Kat, Theorem 1.3] the following q-analogue of Kostant's weight multiplicity formula for χ_λ, that was conjectured by Lusztig in [Lus1]: we have

$$\chi_\lambda = \sum_{\substack{\mu\in\Lambda^{++}\\ \mu\leq\lambda}} K_{\lambda,\mu}(q)\, P_{W_\mu}^{-1} \sum_{w\in W} e^{w\mu} \prod_{i=1}^{\ell} \frac{1-qe^{-wa_i}}{1-e^{-wa_i}}, \tag{5.5.3}$$

where P_{W_μ} is the Poincaré polynomial of the stabilizer W_μ of μ in W.

The partition function formula has led to continuing analysis of the connection between the q-weight multiplicities, functions on nilpotent orbits, filtrations of weight spaces by the kernels of powers of a regular nilpotent element, and degrees in harmonic polynomials (see [JLZ]).

5.5.2 Formal degrees

Let $(\pi, \mathcal{V})$ be a smooth irreducible representation of G. Recall the matrix coefficient π_{v,v^*} defined in (3.5.3). Since π is irreducible, there is a character χ of the centre Z_G of G, called the *central character* of π, such that $\pi(z) = \chi(z)$ for all $z \in Z_G$. We assume that χ is unitary. For every $(v, v^*) \in V \times V^*$, the function $g \mapsto |\pi_{v,v^*}(g)|^2$ is constant on cosets of Z_G, and hence defines a function on G/Z_G.

An irreducible unitary representation with square integrable matrix coefficients is called a *square integrable representation*. Recall that the discrete series of G is the collection of square integrable (mod the center of G) representations. It is also useful to slightly weaken the definition of discrete series, and to define a representation to be in the *essentially discrete series* if it becomes a discrete series representation after twisting by some character of the group.

If f is a function on G, we denote by $\check{f}$ the function defined by $\check{f}(g) := f(g^{-1})$. Let $\mathrm{d}g$ be a fixed Haar measure on G/Z_G. If $(\pi, \mathcal{V})$ is a discrete series representation then there exists a positive real constant $\mathrm{fdeg}(\pi)$, called the *formal degree* of π, such that for all $v_1, v_2 \in \mathcal{V}$, and all $v_1^*, v_2^* \in \mathcal{V}^*$, we have

$$\int_{G/Z_G} \pi_{v_1,v_1^*}(g)\cdot \check{\pi}_{v_2,v_2^*} g)\mathrm{d}g = \frac{v_1^*(v_2)\cdot v_2^*(v_1)}{\mathrm{fdeg}(\pi)} \tag{5.5.4}$$

(see for instance [Ren, IV.3.3]). We define the formal degree of an essentially discrete series representation π as the formal degree of any discrete series representation of G obtained from π by twisting by a character of G.

For an irreducible discrete series representation π of G, the formal degree of π is the value of the Plancherel measure (which depends on the choice of a Haar measure on G) on the singleton set $\{\pi\}$. When G is compact and the Haar measure is chosen so that G has volume 1, then the formal degree is simply the dimension of π. A conjecture of Hiraga–Ichino–Ikeda (see §5.5.5) postulates that the formal degree is equal to the adjoint gamma factor of π evaluated at $s = 0$; they also verified their conjecture for $\mathrm{GL}_n(k)$. A refinement of the formal degree conjecture due to Gross and Reeder [GrRe, Conjecture 8.3] predicts the root number of the adjoint representation.

In practice, it is useful to slightly weaken the definition of a discrete series representation, and to define a representation to be *essentially discrete series* if it becomes discrete series after twisting by some character of the group.

5.5.3 The Plancherel formula

Fix a minimal k-parabolic subgroup $\mathbf{P}_0$ of $\mathbf{G}$ with Levi decomposition $\mathbf{P}_0 = \mathbf{L}_0\mathbf{U}_0$, where $\mathbf{L}_0$ is a minimal k-Levi subgroup of $\mathbf{G}$ and $\mathbf{U}_0$ the unipotent radical of $\mathbf{P}_0$. Let $\mathbf{A}_0$ denote the split component of $\mathbf{L}_0$, that is, the maximal k-split torus in the centre of $\mathbf{L}_0$. Define

$$W_{\mathbf{G}} := W(\mathbf{G}, \mathbf{A}_0) := \mathrm{N}_{\mathbf{G}}(\mathbf{A}_0)/\mathrm{C}_{\mathbf{G}}(\mathbf{A}_0) \tag{5.5.5}$$

to be the Weyl group of $\mathbf{A}_0$ in $\mathbf{G}$, where $\mathrm{N}_{\mathbf{G}}(\mathbf{A}_0)$ and $\mathrm{C}_{\mathbf{G}}(\mathbf{A}_0)$ are the normalizer and the centralizer of $\mathbf{A}_0$ in $\mathbf{G}$, respectively. We denote by Σ_0 the set of roots of $\mathbf{G}$ with respect to $\mathbf{A}_0$, and by Σ_0^+ the set of positive roots determined by the choice of $\mathbf{P}_0$. Let $\Pi_0 \subset \Sigma_0$ be the set of simple roots.

Let $\mathbf{P} \supset \mathbf{P}_0$ be a k-parabolic subgroup of $\mathbf{G}$ with Levi factor $\mathbf{L} = \mathbf{L}_I \supset \mathbf{L}_0$ generated by a subset $I \subset \Pi_0$ and its unipotent radical $\mathbf{U} \subset \mathbf{U}_0$. Let $\mathbf{A}_{\mathbf{L}}$ be the split component of $\mathbf{L}$. We denote by Σ_I the subset of Σ_0 consisting of $\mathbb{Z}$-linear combinations of the roots in I, and set $\Sigma_I^+ := \Sigma_I \cap \Sigma_0^+$. We denote by $\mathbf{P}^- := \mathbf{L}\mathbf{U}^-$ the parabolic subgroup of $\mathbf{G}$ opposite to $\mathbf{P}$, where $\mathbf{U}^-$ is opposite to $\mathbf{U}$.

Let $X^*(\mathbf{L})_k$ denote the group of k-rational characters of $\mathbf{L}$. We set

$$\begin{aligned} \mathfrak{a}_L &:= \mathrm{Hom}(X^*(\mathbf{L})_k, \mathbb{R}) = \mathrm{Hom}(X^*(\mathbf{A}_{\mathbf{L}})_k, \mathbb{R}) \quad \text{and} \\ \mathfrak{a}_L^* &:= X^*(\mathbf{A}_{\mathbf{L}})_K \otimes_{\mathbb{Z}} \mathbb{R} = X^*(\mathbf{L})_k \otimes_{\mathbb{Z}} \mathbb{R}. \end{aligned}$$

We write $\mathfrak{a}_{L,\mathbb{C}}^* := \mathfrak{a}_L \otimes_{\mathbb{R}} \mathbb{C}$. We define the homomorphism $H_L \colon L \longrightarrow \mathfrak{a}_L$ by

$$q^{\langle \chi, H_L(\ell)\rangle} = |\chi(\ell)|_k \quad \text{for all } \chi \in X^*(\mathbf{L})_k \text{ and } \ell \in L. \tag{5.5.6}$$

We fix a maximal special compact subgroup K of G. By using the Iwasawa decomposition $G = PK$, we can extend H_L to G by $q^{\langle \chi, H_L(\ell u k)\rangle} = |\chi(\ell)|_k$.

For $\nu \in \mathfrak{a}_{\mathrm{L},\mathbb{C}}^*$, $\tau \in \mathrm{Irr}_{\mathrm{u}}(L)$, we denote by $(I(\tau, \nu), V(\tau, \nu))$ the normalized parabolically induced representation

$$\mathrm{I}(\tau, \nu) := \mathrm{I}_{L,P}^{G}(\tau \otimes q^{\langle \nu, H_L(\cdot)\rangle} \otimes 1). \tag{5.5.7}$$

We set $U_w := U_0 \cap wUw^{-1}$, and fix a Haar measure $\mathrm{d}u$ on U_w. Given $f \in V(\tau, \nu)$, the *standard intertwining operator* is defined by

$$A(\tau, \nu, w)(g) := \int_{u \in U_w} f(w^{-1}ug)\,\mathrm{d}u \quad \text{for } g \in G. \tag{5.5.8}$$

The integral converges absolutely if $\Re\langle \nu, a^\vee \rangle \gg 0$, for any $a \in \Sigma_0 \backslash I$, where $a^\vee$ is the coroot of a. Moreover, we have

$$A(\tau, \nu, w^{-1})\colon I(\tau, \nu) \to I({}^w\tau, w(\nu)),$$

where ${}^w\tau(\ell_w) := \tau(w^{-1}\ell w)$ for $\ell_w \in wLw^{-1}$. Setting $\overline{U}_w := w^{-1}U_0 w \cap U^-$, we define

$$\gamma_w(G|L) := \int_{\overline{U}_w} q^{\langle 2\rho_P, H_L(\overline{u}) \rangle}\,\mathrm{d}\overline{u}, \tag{5.5.9}$$

where $\mathrm{d}\overline{u}$ is the normalized Haar measure on $\overline{U}_w$ so that the measure of $K_0 \cap \overline{U}_w$ is 1. Here, K_0 is a maximal special parahoric subgroup of G.

For $\nu \in \mathfrak{a}^*_{\mathrm{L},\mathbb{C}}$, $\tau \in \mathrm{Irr}^{\mathrm{u}}(L)$ and $w(I) \subset \Pi_0$, the *Plancherel measure* attached to ν, τ and w for $\mathrm{i}_{L,P}(\tau, \nu)$ is defined to be the non-zero complex number $\mu^L(\tau, \nu, w)$ such that

$$A(\tau, \nu, w) \circ A(w(\tau), w(\nu), w^{-1}) = \mu^L(\tau, \nu, w)^{-1}\gamma_w(G|L)^2. \tag{5.5.10}$$

The Plancherel measure $\mu^L(\tau, \nu, w)$ depends only on τ, ν and w. It is independent of the choices of any Haar measure and of any representative $\widetilde{w}$ of w (see [Sha2, p. 280]). The function

$$\mu^L\colon \tau \mapsto \mu^L(\tau, 0, 1) \tag{5.5.11}$$

is the Harish-Chandra μ-function for L ([Sil, §1] or [Wal, §V.2]).

We denote by $\Sigma(A_L) \subset X^*(A_L)$ the set of non-zero weights occurring in the adjoint representation of A_L on the Lie algebra of G, and by $\Sigma_{\mathrm{r}}(A_L)$ be the set of indivisible elements therein. (Recall that a root a in a root system Σ is called *indivisible* if $\frac{1}{2}a \notin \Sigma$.) For every $a \in \Sigma_{\mathrm{r}}(A_L)$, let $L_a \supset L$ denote the centralizer of $\ker a$ in G (it is a Levi subgroup of G whose semisimple rank is one larger than that of L). We set

$$c(G|L) := \gamma(G|L)^{-1} \prod_{a \in \Sigma_0(P, A_L)} \gamma(L_a|L). \tag{5.5.12}$$

For $\tau \in \mathrm{Irr}^2(L)$, we denote by $\mathrm{fdeg}(\tau)$ the formal degree of τ with respect to the Haar measure $\mathrm{d}l$ on L.

Let $\mathfrak{X}_{\mathrm{u}}(L)$ be the group of unitary unramified characters of L. We define $\mathrm{d}\tau$ to be the measure at $\tau \in \mathrm{Irr}^2(L)$ that is transferred from the normalized Haar measure $\mathrm{d}\chi$ on the quotient $\mathfrak{X}_{\mathrm{u}}(L)/\mathrm{Stab}_{\mathfrak{X}_{\mathrm{u}}}(\tau)$ via the natural isomorphism

$$\mathfrak{X}_{\mathrm{u}}(L)/\mathrm{Stab}_{\mathfrak{X}_{\mathrm{u}}}(\tau) \simeq \{\tau \otimes \chi \,:\, \chi \in \mathfrak{X}_{\mathrm{u}}(L)\}. \tag{5.5.13}$$

The Harish-Chandra Plancherel theorem states that, for any $f \in C_c^\infty(G)$, $f(1)$ is equal to

$$\sum_L c(G|L)^{-2}\gamma(G|L)^{-1}|W_L|^{-1}\int_{\tau\in\mathrm{Irr}^2(L)}\mu^L(\tau)\,\mathrm{fdeg}(\tau)\,\mathrm{trace}(\mathrm{i}_{L,P}^G(\tau))(f)\mathrm{d}\tau,$$

where the sum runs over all k-Levi subgroups L of up to G-conjugacy (see [Wal, Theorem VIII.1.1]).

We set $W(L) := \mathrm{N}_G(L)/L$, where $\mathrm{N}_G(L)$ denotes the normalizer of L in G, and define

$$W^{\mathfrak{s}} := \{n \in \mathrm{N}_G(L) \,:\, {}^n\mathcal{O} \simeq \mathcal{O}\}/L. \tag{5.5.14}$$

The restriction of $\mu = \mu^G$ to $\mathcal{O}$ is a rational, $W^{\mathfrak{s}}$-invariant function on $\mathcal{O}$ (see [Wal, Lemma V.2.1]). It determines a reduced root system [Hei, Proposition 1.3]

$$\Sigma_{\mathfrak{s},\mu} := \left\{a \in \Sigma_{\mathrm{r}}(A_L) \,:\, \mu^{L_a} \text{ has a zero on } \mathcal{O}\right\}. \tag{5.5.15}$$

and the Weyl group $W(\Sigma_{\mathfrak{s},\mu})$ of $\Sigma_{\mathfrak{s},\mu}$ can be identified with the subgroup of $W(L)$ generated by the reflections s_a with $a \in \Sigma_{\mathfrak{s},\mu}$, and as such is a normal subgroup of $W^{\mathfrak{s}}$.

The parabolic subgroup P determines a set of positive roots $\Sigma_{\mathfrak{s},\mu}^+$ and a basis $\Pi_{\mathfrak{s},\mu}$. Let $\ell_{\mathfrak{s}}$ denote the length function on $W(\Sigma_{\mathfrak{s},\mu})$ specified by $\Pi_{\mathfrak{s},\mu}$. Since $W^{\mathfrak{s}}$ acts on $\Sigma_{\mathfrak{s},\mu}$, the function $\ell_{\mathfrak{s}}$ extends naturally to $W^{\mathfrak{s}}$ by

$$\ell_{\mathfrak{s}}(w) := |w(\Sigma_{\mathfrak{s},\mu}^+) \cap -\Sigma_{\mathfrak{s},\mu}^+|. \tag{5.5.16}$$

The set $\Sigma_{\mathfrak{s},\mu}^+$ also determines a subgroup of $W^{\mathfrak{s}}$:

$$R^{\mathfrak{s}} := \left\{w \in W^{\mathfrak{s}} \,:\, w(\Sigma_{\mathfrak{s},\mu}^+) = \Sigma_{\mathfrak{s},\mu}^+\right\} = \left\{w \in W^{\mathfrak{s}} \,:\, \ell_{\mathfrak{s}}(w) = 0\right\}. \tag{5.5.17}$$

By [HumJ, Theorem 1.8], since $W(\Sigma_{\mathfrak{s},\mu})$ acts simply transitively on $\Sigma_{\mathfrak{s},\mu}^+$, we have

$$W^{\mathfrak{s}} = W(\Sigma_{\mathfrak{s},\mu}) \rtimes R^{\mathfrak{s}}. \tag{5.5.18}$$

Let L^1 be the subgroup of L generated by all compact subgroups of L, so that $\mathfrak{X}_{\mathrm{nr}}(L) = \mathrm{Irr}(L/L^1)$. Let σ_1 be an irreducible component of the restriction of σ to L^1. Consider

$$\mathfrak{X}_{\mathrm{nr}}(L,\sigma) := \{\chi \in \mathfrak{X}_{\mathrm{nr}}(L) \,:\, \chi \otimes \sigma \simeq \sigma\}, \tag{5.5.19}$$

which is a finite subgroup of $\mathfrak{X}_{\mathrm{nr}}(L)$, and

$$L_\sigma := \bigcap_{\chi\in\mathfrak{X}_{\mathrm{nr}}(L,\sigma)} \ker\chi, \tag{5.5.20}$$

which has finite index in L^1. We have

$$\operatorname{Irr}(L_\sigma/L^1) \simeq \mathfrak{X}_{\mathrm{nr}}(L)/\mathfrak{X}_{\mathrm{nr}}(L,\sigma) \quad \text{and} \quad \mathbb{C}[L_\sigma/L^1] \simeq \mathbb{C}[\mathfrak{X}_{\mathrm{nr}}(L)/\mathfrak{X}_{\mathrm{nr}}(L,\sigma)]. \tag{5.5.21}$$

Set $(L_\sigma/L^1)^\vee := \operatorname{Hom}_{\mathbb{Z}}(L_\sigma/L^1, \mathbb{Z})$. Composing H_L with the $\mathbb{R}$-linear extension of $H_L(L_\sigma/L^1) \to \mathbb{Z}$ gives an embedding

$$H_L^\vee \colon (L_\sigma/L^1)^\vee \to \mathfrak{a}_L^*. \tag{5.5.22}$$

Let $l_a^\vee$ denote the unique generator of $(L_\sigma \cap L_a)/L^1 \cong \mathbb{Z}$ such that $\operatorname{val}_k(a(l_a^\vee)) > 0$. We define $X_a \in \mathbb{C}(\mathfrak{X}_{\mathrm{nr}}(L)/\mathfrak{X}_{\mathrm{nr}}(L,\sigma))$ by

$$X_a(\chi) := \chi(l_a^\vee) \quad \text{for } \chi \in \mathfrak{X}_{\mathrm{nr}}(L)/\mathfrak{X}_{\mathrm{nr}}(L,\sigma). \tag{5.5.23}$$

We recall the description of the Plancherel measure from [Sil]: for $a \in \Sigma_{\mu,\mathfrak{s}}$ there exist $q_a, q_{a^*} \in \mathbb{R}_{\geq 1}$ and $c'_{s_a} \in \mathbb{R}_{>0}$ such that

$$\mu^{L_a}(\sigma\otimes\cdot) = c'_{s_a} \frac{(1-X_a)(1-X_a^{-1})}{(1-q_a^{-1}X_a)(1-q_a^{-1}X_a^{-1})} \cdot \frac{(1+X_a)(1+X_a^{-1})}{(1+q_{a^*}^{-1}X_a)(1+q_{a^*}^{-1}X_a^{-1})}. \tag{5.5.24}$$

For an explicit computation of the Plancherel formula for G an inner form of $\mathrm{GL}_n(k)$, and how it is related to the Bernstein centre, see [AP1, AP2].

5.5.4 A short excursion through the local Langlands program

The Langlands program envisions deep links between arithmetic and analysis, and uses constructions in arithmetic to predict maps between spaces of functions on different groups. Its core is the construction of a *correspondence* between representations of Galois groups (and related versions) and representations of reductive algebraic groups.

Let k be a non-Archimedean local field, and k' a finite extension of k. Denoting by κ (resp. κ')) the residue field of k (resp. k'), we set $f := [\kappa' : \kappa]$. Let e be the integer such that $\varpi_k \mathfrak{o}_{k'} = \mathfrak{p}_{k'}^e$, where $\mathfrak{o}_k$, and $\mathfrak{o}_{k'}$ are the rings of integers of k and k', respectively. The integers f and e are the *residue degree* and *ramification degree* of k' over k. The extension k'/k is said to be *unramified* if $e = 1$.

We fix a separable algebraic closure $\overline{k}$ of k, and set

$$\Gamma_k := \operatorname{Gal}(\overline{k}/k) := \varprojlim_{k'} \operatorname{Gal}(k'/k), \tag{5.5.25}$$

where k'/k ranges over finite Galois extensions with $k' \subset \overline{k}$.

Finite unramified extensions

Let k_m be the unramified extension of k of degree m. It is Galois and $\mathrm{Gal}(k_m/k)$ is cyclic. There is a unique element ϕ_m of $\mathrm{Gal}(k_m/k)$ which acts on $\kappa_m \simeq \mathbb{F}_{q^m}$ as $x \mapsto x^q$, where κ_m is the residue field of k_m. Set $\Phi_m := \phi_m^{-1}$. Then $\Phi_m \mapsto 1$ defines a canonical isomorphism $\mathrm{Gal}(k_m/k) \xrightarrow{\sim} \mathbb{Z}/m\mathbb{Z}$.

Maximal unramified extensions

The composite of all the k_m is the maximal unramified extension of k, which we will denote by k_{nr}. There is a canonical isomorphism of topological groups

$$\mathrm{Gal}(k_{\mathrm{nr}}/k) \simeq \varprojlim_{m\geq 1} \mathbb{Z}/m\mathbb{Z} =: \widehat{\mathbb{Z}}, \tag{5.5.26}$$

and we denote by $\Phi_k \in \mathrm{Gal}(k_{\mathrm{nr}}/k)$ the unique element that acts on k_m as Φ_m, for all m. An element of Γ_k is called a *geometric Frobenius* if its image in $\mathrm{Gal}(k_{\mathrm{nr}}/k)$ is Φ_k.

The absolute Weil group

The group $I_k := \mathrm{Gal}(\overline{k}/k_{\mathrm{nr}})$ is called the *inertia group* of k. We have an exact sequence of topological groups

$$1 \longrightarrow I_k \longrightarrow \Gamma_k \longrightarrow \widehat{\mathbb{Z}} \longrightarrow 0. \tag{5.5.27}$$

Let ${}_aW_k$ be the inverse image in Γ_k of the cyclic subgroup $\langle \Phi_k \rangle$ of $\mathrm{Gal}(k_{\mathrm{nr}}/k)$ generated by Φ_k. It is a dense subgroup of Γ_k and fits into an exact sequence of abstract groups:

$$1 \longrightarrow I_F \longrightarrow_a W_k \longrightarrow \mathbb{Z} \longrightarrow 0. \tag{5.5.28}$$

The (absolute) Weil group of k is defined to be the topological group, with underlying abstract group ${}_aW_k$, so that I_k is an open subgroup of W_k, and the topology on I_k, as subspace of W_k, coincides with its natural topology as $\mathrm{Gal}(\overline{k}/F_{\mathrm{nr}}) \subset \Gamma_k$ (see [BH] or [Tat] for more details).

Local constants

Let ψ be a fixed nontrivial additive character of k. As in [Tat], we attach to a finite-dimensional complex representation (r, V) of W_k epsilon factors $\varepsilon(s, r, \psi)$ and L-functions $L(s, r)$, where $s \in \mathbb{C}$ is a complex variable. The L-function $L(s, r)$ is a meromorphic function of s which only depends on the semisimplification of V.

It is defined as

$$L(s,r) := \det\left(1 - q^{-s} r(\mathrm{Frob}_k)\,|_{V^{I_k}}\right)^{-1}, \tag{5.5.29}$$

where I_k is the inertia group of k and $\mathrm{Frob}_k \in W_k$ is a Frobenius element. The L-function satisfies additivity and inductivity, i.e., if k' is a subfield of k and r a finite-dimensional complex representation of W_k, then we have

$$L(s, \mathrm{Ind}_{W_k}^{W_{k'}}(r)) = L(s,r). \tag{5.5.30}$$

We define

$$\gamma(s,r,\psi) := \varepsilon(s,\pi,r,\psi)\cdot \frac{L(1-s,r)}{L(s,\pi,r)}. \tag{5.5.31}$$

Recently, Oi, Sakamoto and Tamori computed in [OST] the Hecke action of a certain test function on the space of an unramified principal series of G. As a consequence, they obtained a new expression of the local L-functions of unramified representations.

The Langlands dual group of G

The Langlands dual group of G is the complex Lie group $G^\vee$ with root datum dual to that of G. Here are a few examples.

G	Dynkin diagram	$G^\vee$
$\mathrm{GL}_n(\mathbb{Q}_p)$		$\mathrm{GL}_n(\mathbb{C})$
$\mathrm{SL}_n(\mathbb{Q}_p)$		$\mathrm{PSL}_n(\mathbb{C})$
$\mathrm{PGL}_n(\mathbb{Q}_p)$		$\mathrm{SL}_n(\mathbb{C})$
$\mathrm{Sp}_{2n}(\mathbb{Q}_p)$		$\mathrm{SO}_{2n+1}(\mathbb{C})$
$\mathrm{SO}_{2n+1}(\mathbb{Q}_p)$		$\mathrm{Sp}_{2n}(\mathbb{C})$
$\mathrm{SO}_{2n}(\mathbb{Q}_p)$		$\mathrm{SO}_{2n}(\mathbb{C})$
$\mathrm{G}_2(\mathbb{Q}_p)$		$\mathrm{G}_2(\mathbb{C})$

Let $Z_{G^\vee}$ denote the center of $G^\vee$. Let Ad denote the adjoint representation of ${}^LG$ on $\mathrm{Lie}(G^\vee)/\mathrm{Lie}({}^LZ_{G^\vee})$. The gamma factor $\gamma(s,\pi,\mathrm{Ad},\psi)$ is called the *adjoint γ-factor* of π.

Enhanced L-parameters

We denote by $W_k' := W_k \times \mathrm{SL}_2(\mathbb{C})$ the Weil–Deligne group of k. We suppose that G is k-split. A *Langlands parameter* – or L-parameter – is a morphism $\varphi\colon W_k' \to G^\vee$ such that

- $\varphi|_{\mathrm{SL}_2(\mathbb{C})}$ is a morphism of algebraic groups,
- $\varphi(w)$ is a semisimple element of $G^\vee$, for any $w \in W_k$.

The *local Langlands correspondence* predicts a surjective map, satisfying several properties,

$$\left\{\begin{matrix}\text{irred. smooth}\\ \text{repres. } \pi \text{ of } G\end{matrix}\right\}/\text{iso.} \longrightarrow \left\{\begin{matrix}L\text{-parameters}\\ \varphi_\pi : W_F \times \mathrm{SL}_2(\mathbb{C}) \to {}^LG\end{matrix}\right\}/G^\vee\text{-conj.} \tag{5.5.32}$$

with finite fibers, called *L-packets*.

In order to obtain a bijection between the group side and the Galois side, the conjectural map $\mathcal{L}$ was later enhanced: on the Galois side, one considers enhanced L-parameters (φ_π, ρ_π). For φ a given L-parameter, we denote by $\mathrm{C}_{G^\vee}(\varphi)$ the centralizer in $G^\vee$ of $\varphi(W_k)$, a (possibly disconnected) reductive group, and define

$$S_\varphi := \mathrm{C}_{G^\vee}(\varphi)/\mathrm{Z}_{G^\vee} \cdot \mathrm{C}_{G^\vee}(\varphi)^\circ. \tag{5.5.33}$$

By definition, an *enhancement* of φ is an irreducible representation ρ of S_φ. Pairs (φ, ρ) are called *enhanced L-parameters* for G. Let $G^\vee$ act on the set of enhanced L-parameters via

$$g \cdot (\varphi, \rho) = (g\varphi g^{-1}, g \cdot \rho). \tag{5.5.34}$$

We denote by $\Phi_{\mathrm{e}}(G)$ the set of $G^\vee$-conjugacy classes of enhanced L-parameters. This set admits a Bernstein-type decomposition (see [AMS] and [Au1]).

A bijective LLC has been constructed in particular in the following cases:

- $G = k^\times = \mathrm{GL}_1(k)$: Class field theory (first half of the 20th century);
- $G = \mathrm{GL}_n(k)$: Laumon, Rapoport and Stuhler [LRS], when k has positive characteristic, [HT], [Hen], [Scho], when k has arbitrary characteristic;
- $G = \mathrm{SL}_n(F)$ (and its inner twists): Hiraga and Saito [HS], when k has characteristic zero, [ABPS2], when k has positive characteristic;
- $G = \mathrm{Sp}_{2n}(k), \mathrm{SO}_{2n+1}(k)$: Arthur [Art], when k has zero characteristic;
- $G = \mathrm{G}_2(k)$: two independent constructions in [AX1, AX2] and [GS];
- for all the unipotent representations of an arbitrary p-adic group G in [Lus3], [Lus4], extended in [FOS2], [Op6], [So1].

5.5.5 The Hiraga–Ichino–Ikeda conjecture

Let k be a non-Archimedean local field. In [HII], Hiraga, Ichino and Ikeda conjectured an explicit expression for the Plancherel density of $G = \mathbf{G}(k)$, in terms of Langlands parameters. For an essentially discrete series representation π in an L-packet $\Pi_\varphi(G)$ attached to a (discrete) Langlands parameter $\varphi\colon L_F \to {}^LG$, enhanced with a local system $\rho_\pi \in \mathrm{Irr}(S_\varphi, \chi_G)$ on the $G^\vee$-orbit of φ the conjecture reduces to the equality

$$\mathrm{fdeg}(\pi) = \frac{\dim(\rho_\pi)}{|\, S_\varphi|}\, |\gamma(0, \mathrm{Ad} \circ \varphi, \psi)\,| \,. \tag{5.5.35}$$

Hiraga, Ichino, and Ikeda proved the conjecture for $k = \mathbb{R}$ [HII]. For k non-Archimedean, it has been proved in several cases (see notably [Beu], [Oh], [FOS2], [Kak]) but is still open in general.

5.5.6 Spherical functions on p-adic homogeneous spaces

The classical analysis on symmetric spaces and their spherical functions due to Harish-Chandra [8] and Helgason [Hel2] is governed by associated root systems. Starting in the late 1980s, there were important developments by Heckman and Opdam, Dunkl and Cherednik, who established the foundations of harmonic analysis related to root systems beyond symmetric spaces, see [Op2], where spherical functions are replaced by general families of multivariate orthogonal polynomials and hypergeometric functions which depend on continuous parameters. A key ingredient in their study are commuting differential-reflection operators, called Dunkl operators, which replace the invariant differential operators on a Riemannian symmetric space. The article [Gan] provides a survey of several questions and results on spherical functions, mainly following Harish-Chandra's works. Sections 2 and 3 of [Gan] are rather parallel to the p-adic situation, with K a maximal compact subgroup of G, which guarantees that the convolution algebra of continuous compactly supported K-biinvariant functions is commutative.

Let $\mathbf{G}$ be a reductive linear algebraic group defined over a p-adic field k, and $\mathbf{X}$ an affine algebraic variety defined over k which is $\mathbf{G}$-homogeneous. Put $G := \mathbf{G}(k)$ and $X := \mathbf{X}(k)$. Then the spherical Hecke algebra $\mathcal{H}(G, K)$ of G with respect to a compact subgroup K of G acts by convolution product on the space $\mathcal{C}^\infty(K\backslash X)$ of K-invariant $\mathbb{C}$-valued functions on X. A non-zero function in $\mathcal{C}^\infty(K\backslash X)$ is called a *spherical function on* X if it is a common $\mathcal{H}(G, K)$-eigenfunction.

Spherical functions on homogeneous spaces have also been studied as spherical vectors of distinguished models, Shalika functions and Whittaker–Shintani functions, and are closely related to theory of automorphic forms and to representation theory. When G and X are defined over $\mathbb{Q}$, spherical functions appear in local factors of global objects, e.g. Rankin–Selberg convolutions and Eisenstein series.

In [Hi5], Hironaka obtained a general expression of spherical functions on X in terms of the rational functions determined by the group G and the matrix appearing in the functional equations of spherical functions, and constructed a unified method to obtain explicit functional equations of spherical functions on X.

Previously, Hironaka and Sato developed in [HS1, HS2] a theory of spherical functions on spaces of nondegenerate alternating matrices over a p-adic field k, and used it to calculate local densities of alternating forms. For a positive integer n, let X be the set of nondegenerate alternating matrices of size $2n$ with entries in k, and set $G := \mathrm{GL}_{2n}(k)$ and $K := \mathrm{GL}_{2n}(\mathfrak{o})$. The group G acts on X via $g \cdot x := gx^t g$ for $x \in X$ and $g \in G$. The action is transitive and X is identified with the symmetric homogeneous space $\mathrm{GL}_{2n}(k)/\mathrm{Sp}_{2n}(k)$.

Let $\mathcal{S}(K\backslash X)$ be the subspace of $\mathcal{C}^\infty(K\backslash X)$ consisting of all compactly supported functions in $\mathcal{C}^\infty(K\backslash X)$. We define an action of $\mathcal{H}(G, K)$ on $\mathcal{C}^\infty(K\backslash X)$ via the convolution product. Then $\mathcal{S}(K\backslash X)$ is an $\mathcal{H}(G, K)$-submodule of $\mathcal{C}^\infty(K\backslash X)$. Let S_n denote the symmetric group in n letters. An isomorphism of $\mathcal{S}(K\backslash X)$ to $\mathbb{C}[x_1^{\pm 1}, \dots, x_n^{\pm 1}]^{S_n}$, as $\mathcal{H}(G, K)$-modules, is constructed in [HS1, Theorem 1]. The spherical functions are parametrized by $\mathbb{C}^n/S_n$, and can be represented by an integral analogous to the integral expressions of the zonal spherical functions on p-adic reductive groups (see [HS1, Theorem 2]). The Plancherel measure is computed in [HS1, Theorem 4] by similar arguments as in Section 5.1.

In [Hi1, Hi2, Hi3, Hi4], Hironaka established an analogous theory of spherical functions and a spherical transform for the space of nondegenerate hermitian forms on k, and applied it to calculate local densities of unramified hermitian forms. The case of quaternion Hermitian forms is studied in [Hi6], and that of unitary Hermitian forms in [HY].

5.5.7 Spherical varieties

The description of irreducible representations of a group G can be seen as a problem in harmonic analysis; namely, decomposing a suitable space of functions on G into irreducible subspaces for the action of $G \times G$ by left and right multiplication. For a split p-adic reductive group $\mathbf{G}$ over k, unramified irreducible smooth representations are in bijection with semisimple conjugacy classes in the Langlands dual group $G^\vee$ of G.

Let $\mathbf{G}$ be an algebraic group over an arbitrary field k in characteristic zero. A $\mathbf{G}$-*variety* (over k) is defined to be a geometrically integral and separated k-scheme of finite type with an algebraic action of $\mathbf{G}$ over k. A $\mathbf{G}$-variety $\mathbf{X}$ is called *homogeneous* if $\mathbf{G}(\overline{k})$ acts transitively on $\mathbf{X}(\overline{k})$; then $\mathbf{X}$ is automatically nonsingular. If $\mathbf{X}$ has a point x over k, its stabilizer $\mathrm{Stab}_{\mathbf{G}}(x)$ is a subgroup over k and $\mathbf{X} \cong \mathrm{Stab}_{\mathbf{G}}(x)\backslash\mathbf{G}$, the geometric quotient of $\mathbf{G}$ by $\mathrm{Stab}_{\mathbf{G}}(x)$. Conversely, for any closed subgroup $\mathbf{H}$ of $\mathbf{G}$ the geometric quotient $\mathbf{H}\backslash\mathbf{G}$ is a homogeneous variety under the action of $\mathbf{G}$.

We suppose that $\mathbf{G}$ is a reductive group over a field k. Let $\mathbf{B}$ be a Borel subgroup of $\mathbf{G}$. A normal $\mathbf{G}$-variety $\mathbf{X}$ over k (not necessarily homogeneous) is called *spherical* if $\mathbf{B}(\overline{k})$ has a Zariski open orbit on $\mathbf{X}(\overline{k})$. This is equivalent to the existence of finitely many $\mathbf{B}(\overline{k})$-orbits [Bri, Vin].

From now on we assume that $\mathbf{X}$ is quasi-affine. Every spherical variety $\mathbf{X}$ contains a finite number of $\mathbf{G}$-orbits, and each of them is also spherical (see [Kn]).

We assume that $\mathbf{G}$ is k-split and the characteristic of k equals 0. Let $\mathbf{B}$ be a Borel subgroup of $\mathbf{G}$, with a maximal torus $\mathbf{T} \subset \mathbf{B}$. Let $\mathbf{U}$ denote the unipotent radical of $\mathbf{B}$, and let G, B, T and U be the k-rational points of $\mathbf{G}$, $\mathbf{B}$, $\mathbf{T}$ and $\mathbf{U}$, respectively. The complex torus $\mathfrak{X}_{\mathrm{nr}}(T)$ of the unramified characters of T (considered simultaneously as characters of B via $T = B/U$) is canonically isomorphic to $T^\vee$, the maximal torus in the Langlands dual group $G^\vee$ of G. In [Sa2], Sakellaridis extended this description to an arbitrary spherical variety $\mathbf{X}$ of $\mathbf{G}$.

In [Sa3], Sakellaridis generalized and unified all previously known results on spherical functions over local fields to the case of *homogeneous spherical varieties* for a split reductive group. Some technical conditions are imposed (see [Sa3, §1.7]); in particular, X is assumed to be quasi-affine. A suitable maximal torus $\mathbf{A}$ of $\mathbf{G}$ is chosen and the $\mathbf{A}$-orbit of x is identified with a (torus) quotient $\mathbf{A_X}$. The k-points of $\mathbf{X}$ are denoted by X, and the group $\mathbf{G}(\mathfrak{o})$ of $\mathfrak{o}$-points of $\mathbf{G}$, which is a hyperspecial maximal compact subgroup of G, is denoted by K. Consider the right regular representation of G on $\mathcal{V} := \mathcal{C}^\infty(X)$ (in the case where G has a unique orbit on X, this is just the induced representation $\mathrm{Ind}_H^G(1)$). The unramified (or spherical) Hecke algebra $\mathcal{H}_\mathbb{Z}(G, K)$ acts on $\mathcal{V}$, and the goal of [Sa3] is to compute an explicit formula for its eigenvectors expressed in terms of the geometry of $\mathbf{X}$. Moreover, the Plancherel formula for unramified functions on X is computed in [Sa3, Theorem 9.0.1]. It is known that there exists an essentially unique decomposition of $L^2(X)$ as a direct integral of irreducible, unitary representations.

The study of period integrals of automorphic forms has revealed relationships between the non-vanishing of certain period integrals of automorphic forms, on one hand, and functorial lifts, on the other, and also between the values of these period integrals and L-functions or special values thereof. In the context of the Geometric Langlands program, the "dual group" attached by Gaitsgory and Nadler [GaNa] to any spherical variety X was shown in [Sa2] to be related to unramified representations in the spectrum of X and is, more generally, conjectured in [SV] to describe X-distinguished representations, in a certain sense.

In the spirit of the Langlands philosophy, in [SV], Sakellaridis and Venkatesh associated to the spherical variety X the following dual data: a Langlands dual group $X^\vee$, a map $\iota_X \colon X^\vee \times \mathrm{SL}_2(\mathbb{C}) \to G^\vee$ and a (graded) finite-dimensional (typically) symplectic representation V_X of $X^\vee$. The representation V_X is the main ingredient allowing one to form the automorphic L-function which controls the relevant period. Ben-Zvi, Sakellaridis and Venkatesh proposed in [BZSV] a broader framework for the relative Langlands program. The basic objects considered by the relative Langlands program should then be a class of Hamiltonian G-varieties called *hyperspherical varieties*.

Bibliography

[0] M.F. Atiyah and I.G. Macdonald, *Introduction to Commutative Algebra*. Addison–Wesley (1969).

[1] N. Bourbaki, *Groupes et Algèbres de Lie*, Chapitres 4, 5 et 6, Hermann (1968).

[2] F. Bruhat and J. Tits, 4 notes in *Comptes Rendus* vol. **263** (1966), 598–601; 766–768; 822–825; 867–869.

[3] F. Bruhat and J. Tits, "Groupes algébriques simples sur un corps local", in *Proceedings of a Conference on Local fields*, Springer (1967).

[4] J. Dieudonné, *Treatise on Analysis*, vol. II, Academic Press (1967).

[5] S.G. Gindikin and F.I. Karpelevič, "Plancherel measure of Riemannian symmetric spaces of non-positive curvature", *Soviet Math.* **3** (1962), 962–965.

[6] R. Godement, "Introduction aux travaux de A. Selberg", *Sém. Bourbaki* (1957), Exp. **144**.

[7] R. Godement, "A theory of spherical functions I", *Trans. A.M.S.* vol. **73** (1952), 496–556.

[8] Harish-Chandra, "Spherical functions on a semi-simple Lie group I", *Amer. J. Math.* **80** (1958), 241–310.

[9] Harish-Chandra, "Discrete series for semi-simple Lie groups II", *Acta. Math.* **116** (1966), 1–112.

[10] S. Helgason, *Differential Geometry and Symmetric Spaces*, Academic Press (1962).

[11] S. Helgason and K. Johnson, "The bounded spherical functions on symmetric spaces", *Advances in Math.* **3** (1969), 586–593.

[12] I.G. Macdonald, "Spherical functions on a p-adic Chevalley group", *Bull. A.M.S.* **74** (1968), 520–525.

[13] H. Matsumoto, "Functions sphériques sur un groupe semi-simple p-adique", *CRAS* **269** (1969), 829–832.

[14] I. Satake, "Theory of spherical functions on reductive algebraic groups over p-adic fields", *Publ. Math. I.H.E.S.* **18** (1963), 1–69.

[15] R. Steinberg, *Lectures on Chevalley groups*, Yale (1967).

I. G. Macdonald and A.-M. Aubert, *Spherical Functions on a Group of p-adic Type*,
Lecture Notes in Mathematics 2392,
https://doi.org/10.1007/978-3-032-15671-6

[16] T. Tamagawa, "On Selberg's trace formula", *J. Fac. Sci. Univ. of Tokyo* **8** (1960), 363–386.

[17] J.T. Tate, "Fourier analysis in number fields and Hecke's zeta-functions", reproduced in *Algebraic Number Theory* (eds. J.W.S. Cassels and A. Fröhlich), Academic Press (1967).

Additional Bibliography

[AB] P. Abramenko and K.S. Brown, *Buildings: Theory and Applications*, Springer, Graduate Texts in Mathematics, **248**, 2008.

[AD] J. Adler and S. DeBacker, "Some applications of Bruhat–Tits theory to harmonic analysis on the Lie algebra of a reductive p-adic group", *Michigan Math. J.* **50**:2 (2002), 263–286.

[AST] J.-P. Anker, Br. Schapira, B. Trojan, "Sharp estimates for distinguished random walks on affine buildings of type $\widetilde{A}_r$", *Indagationes Mathematicae*, 2024.

[Art] J. Arthur, *The endoscopic classification of representations: orthogonal and symplectic groups*, Colloquium Publications **61**, Amer. Math. Soc., 2013.

[AG] A. Ash and D. Ginzburg, "p-adic L-functions for $\mathrm{GL}(2n)$", *Invent. Math.* **116** (1994), 27–73.

[Au1] A.-M. Aubert, "Local Langlands and Springer correspondences", in *Representations of Reductive p-adic Groups*, A.-M. Aubert, M. Mishra, A. Roche, S. Spallone (Eds.), Progress in Mathematics **328**, Birkhäuser, 2019, pp. 1–37.

[Au2] A.-M. Aubert, "Bruhat–Tits buildings, representations of p-adic groups and Langlands correspondence", *J. Algebra* **656** (2024), 47–76 (Volume dedicated to Jacques Tits).

[Au3] A.-M. Aubert, "Representation theory of p-adic reductive groups", in *K-Theory and Representation Theory*, eds. H. Şengün and R. Plymen, LMS Textbook Series **494**, 2024.

[ABPS1] A.-M. Aubert, P.F. Baum, R.J. Plymen, and M. Solleveld, "Depth and the local Langlands correspondence", in Arbeitstagung Bonn 2013, In Memory of Friedrich Hirzebruch, eds. W. Ballmann, C. Blohmann, G. Faltings, P. Teichner, and D. Zagier pp. 17–41, Progr. Math. **319**, Birkhäuser/Springer, 2016.

I. G. Macdonald and A.-M. Aubert, *Spherical Functions on a Group of p-adic Type*, Lecture Notes in Mathematics 2392,
https://doi.org/10.1007/978-3-032-15671-6

[ABPS2] A.-M. Aubert, P.F. Baum, R.J. Plymen, and M. Solleveld,"The local Langlands correspondence for inner forms of SL_n", *Res Math Sci* **3**:32 (2016).

[AMS] A.-M. Aubert, A. Moussaoui and M. Solleveld, "Generalizations of the Springer correspondence and cuspidal Langlands parameters", *Manuscripta Math.* **157** (2018), no. 1-2, 121–192.

[AP1] A.-M. Aubert and R.J. Plymen, "Plancherel measure for $GL(n, F)$ and $GL(m, D)$: explicit formulas and Bernstein decomposition", *J. Number Theory* **112**:1 (2005), 26–66.

[AP2] A.-M. Aubert and R.J. Plymen, "Comparison of the depths on both sides of the local Langlands correspondence for Weil-restricted groups", *J. Number Theory* **233** (2022), 24–58.

[AX1] A.-M. Aubert and Y. Xu, "Hecke algebras for p-adic reductive groups and local Langlands correspondences for Bernstein blocks", *Adv. Math.* **436** (2024), Paper No. 109384, 45 pp.

[AX2] A.-M. Aubert and Y. Xu, "The explicit local Langlands correspondence for G_2", arXiv:2208.12391, 2022.

[BI] E. Bannai and T. Ito, *Algebraic Combinatorics I. Association Schemes*, The Benjamin/Cummings Publishing Co., CA, 1984.

[BGR] N. Bardy-Panse, S. Gaussent, and G. Rousseau, "Macdonald's formula for Kac–Moody groups over local fields", *Proc. Lond. Math. Soc.* (3) **119**:1 (2019), 135–175.

[Bar] V. Bargmann, "Irreducible unitary representations of the Lorentz group", *Ann. of Math.* **48** (1947), 568–640.

[BR] P. Baumann and S. Riche, "Notes on the geometric Satake equivalence", Lecture Notes in Math. **2221**, CIRM Jean-Morlet Ser. Springer, Cham, 2018, 1–134.

[BZSV] D. Ben-Zvi, Y. Sakellaridis, and A. Venkatesh, "Relative Langlands Duality", preprint. Available at `https://www.math.ias.edu/~akshay/research/BZSVpaperV1.pdf`.

[Be1] J. Bernstein, "Le "centre" de Bernstein", in: *Representations of reductive groups over a local field*, Travaux en Cours, 1–32, Ed. P. Deligne, Hermann, Paris, 1984.

[Be2] J. Bernstein, "Draft of: Representations of p-adic groups", Lectures by Joseph Bernstein, Harvard University, Fall 1992, Notes by Karl E. Rumelhart. `http://www.math.tau.ac.il/~bernstei/Unpublished_texts/unpublished_texts/Bernstein93new-harv.lect.from-chic.pdf`

[Beu] R. Beuzart-Plessis "Plancherel formula for and applications to the Ichino–Ikeda and formal degree conjectures for unitary groups", *Invent. math.* **225** (2021), 159–297.

[Bo1] A. Borel, *Essays in the History of Lie Groups and Algebraic Groups*, History of Mathematics **21**, Amer. Math. Soc. 2001.

[Bo2] A. Borel, "Automorphic L-functions", Proceedings of Symposia in Pure Mathematics, Vol. 33 (1979), part 2, 27–61.

[BrKa] A. Braverman and D. Kazhdan, "The spherical Hecke algebra for affine Kac–Moody groups I", *Ann. of Math. (2)* **174**:3 (2011), 1603–1642.

[BKP] A. Braverman, D. Kazhdan and M. Patnaik, "Iwahori–Hecke algebras for p-adic loop group", *Invent. Math.* **204** (2016), 347–442.

[Bri] M. Brion, "Quelques propriétés des espaces homogènes sphériques", *Manuscripta Math.* **55** (1986), 191–198.

[Br] P. Broussous, "Acyclicity of Schneider and Stuhler's coefficient systems: another approach in the level 0 case", *J. Algebra* **279** (2004), 737–748.

[Bru1] F. Bruhat, "Sur les représentations induites des groupes de Lie", *Bull. Soc. Math. France* **84** (1956), 97–205.

[Bru2] F. Bruhat, "Distributions sur un groupe localement compact et applications à l'étude des représentations des groupes p-adiques", *Bull. Soc. Math. France* **89** (1961), 43–75.

[Bru3] F. Bruhat, "Sur une classe du sous-groupes compacts maximaux des groupes de Chevalley sur un corps $\mathfrak{p}$-adique", *Inst. Hautes Études Sci. Publ. Math.* **23** (1964), 45–74.

[Bru4] F. Bruhat, "$\mathfrak{p}$-adic groups", in: *Algebraic Groups and Discontinuous Subgroups* (Proc. Sympos. Pure Math., Boulder, Colo., 1965), pp. 63–70 Amer. Math. Soci., Providence, RI, 1966.

[Bru5] F. Bruhat, "Sous-groupes compacts maximaux des groupes semi-simples $\mathfrak{p}$-adiques" , in *Les Tendances Géom. en Algèbre et Théorie des Nombres*, pp. 43–53 Colloq. Internat. CNRS, No. **143**, Éditions du Centre National de la Recherche Scientifique (CNRS), Paris, 1966.

[BT1] F. Bruhat and J. Tits, "Groupes réductifs sur un corps local: Donnée radicielle valuée", *Inst. Haut. Études Sci. Publ. Math.* (1972), 5–251.

[BT2] F. Bruhat and J. Tits, "Groupes réductifs sur un corps local II: Schémas en groupes. Existence d'une donnée radicielle valuée", *Inst. Haut. Études Sci. Publ. Math.* **60** (1984), 197–376.

[BT3] F. Bruhat and J. Tits, "Groupes algébriques sur un corps local Chapitre III. Compléments et applications à la cohomologie galoisienne", *J. Fac. Sci. Univ. Tokyo Sect. IA, Math.* **34** (1987), 671–698.

[BH] C.J. Bushnell and G. Henniart, *The local Langlands conjecture for* $\mathrm{GL}(2)$, Grundlehren der mathematischen Wissenschaften, **335**, Springer-Verlag, Berlin, 2006, xii+347pp.

[Car] P. Cartier, "Representations of p-adic groups: a survey", in: *Automorphic forms, representations and L-functions* (Proc. Sympos. Pure Math., Oregon State Univ., Corvallis, Ore., 1977), Part 1, Proc. Sympos. Pure Math., XXXIII, 111–155. Amer. Math. Soc., Providence, R.I., 1979.

[Cas] W. Casselman, "The unramified principal series of p-adic groups. I. The spherical function", *Compositio Mathematica* **40**:4 (1980), 387–406.

[CS] W. Casselman and J. Shalika, "The unramified principal series of p-adic groups. II. The Whittaker function", *Compositio Math.* **41**:2 (1980), 207–231.

[CCH] W. Casselman, J. Cely, and T. Hales, "The spherical Hecke algebra, partition functions, and motivic integration", *Trans. Amer. Math. Soc.* **371**:9 (2019), 6169–6212.

[Ca1] R. Cass, "Central elements in affine mod p Hecke algebras via perverse $\mathbb{F}_p$ -sheaves", *Compos. Math.* **157**:10 (2021), 2215–2241.

[Ca2] R. Cass, "Perverse $\mathbb{F}_p$-sheaves on the affine Grassmannian", *J. Reine Angew. Math.* **785** (2022), 219–272.

[CR] L. Colmenarejo and A. Ram, "c-functions and Macdonald polynomials", *J. Algebra* **655** (2024), 163–222.

[DB] S. DeBacker, "Some applications of Bruhat–Tits theory to harmonic analysis on a reductive p-adic group", *Michigan Math. J.* **50**:2 (2002), 241–261.

[Die] J.F. van Diejen, "On the Plancherel formula for the (discrete) Laplacian in a Weyl chamber with repulsive boundary conditions at the walls", *Ann. Henri Poincaré* **5** (2004), 135–168, Birkhäuser Verlag, Basel.

[DE] J.F. van Diejen and E. Emsiz, "Unitary representations of affine Hecke algebras related to Macdonald spherical functions", *J. Algebra* **354** (2012), 180–210.

[Dix] J. Dixmier, *Les C^*-algèbres et leurs représentations*, Cahiers scientifiques XXIX, Gauthier-Villars, Paris 1969.

[Ehr] C. Ehresmann, "Sur la topologie de certains espaces homogènes", *Ann. Math.* **35** (1934), 396–443.

[FOS1] Y. Feng, E. Opdam, M. Solleveld, "Supercuspidal unipotent representations: L-packets and formal degrees", *J. Éc. polytech. Math.* **7** (2020), 1133–1193.

[FOS2] Y. Feng, E. Opdam, M. Solleveld, "On formal degrees of unipotent representations", *J. Inst. Math. Jussieu* **21**:6 (2022), 1947–1999.

[FM] T. Finis, J. Matz, *On the asymptotics of Hecke operators for reductive groups*, *Math. Ann.* **380** (2021), 1037–1104.

[Fla] D. Flath, "Decomposition of representations into tensor products", in "Automorphic forms, representations and L-functions", Part 1, Proc. Sympos. Pure Math., XXXIII, pages 179–183. Amer. Math. Soc., Providence, R.I., 1979.

[FGKV] E. Frenkel, D. Gaitsgory, D. Kazhdan, and K. Vilonen, "Geometric realization of Whittaker functions and the Langlands conjecture", *J. Amer. Math. Soc.* **11**:2 (1998), 451–484.

[GAJV] M.-P. Gomez Aparicio, P. Julg, and A. Valette, "The Baum–Connes conjecture: an extended survey", in *Advances in Noncommutative Geometry: On the Occasion of Alain Connes' 70th Birthday*, 127–244. Springer International Publishing, 2019.

[GS] W.T. Gan and G. Savin, "The local Langlands conjecture for G_2", *Forum of Mathematics, Pi* **11** (2023), 1–42.

[GaNa] D. Gaitsgory and D. Nadler, "Spherical varieties and Langlands duality", *Mosc. Math. J.* **10**:1 (2010), 65–137.

[Gan] R. Gangolli, "Spherical functions on semisimple Lie groups. Symmetric spaces" (Short Courses, Washington Univ., St. Louis, Mo., 1969–1970), pp. 41–92. Pure and Appl. Math., Vol. **8**, Dekker, New York, 1972.

[Gar] P. Garrett, *Buildings and Classical Groups*, Chapman & Hall, 1997.

[GaRo1] S. Gaussent and G. Rousseau, "Kac–Moody groups, hovels and Littelmann paths", *Ann. Inst. Fourier (Grenoble)* **58**:7 (2008), 2605–2657.

[GaRo2] S. Gaussent and G. Rousseau, "Spherical Hecke algebras for Kac–Moody groups over local fields", *Ann. of Math. (2)* **180**:3 (2014), 1051–1087.

[Gel] I.M. Gel'fand, "Spherical functions in symmetric Riemann spaces" (Russian), *Doklady Akad. Nauk SSSR (N.S.)* **70** (1950), 5–8.

[GeNa] I.M. Gel'fand and M.A. Naĭmark, "Unitary representations of the Lorentz group" (Russian), *Izv. Akad. Nauk SSSR Ser. Mat.* **11** (1947), 411–504.

[Gi] S. Gindikin, "Harish-Chandra's c-function; 50 years later", *Ann. Fac. Sci. Toulouse Math.* (6) **25** (2016), 385–402.

[Gr] B. Gross, "On the Satake isomorphism", in *Galois representations in arithmetic algebraic geometry* (Durham, 1996), London Math. Soc. Lecture Note Ser., vol. **254**, Cambridge University Press, pp. 223–237.

[GrRe] B. Gross and M. Reeder, "Arithmetic invariants of discrete Langlands parameters", *Duke Math. J.* **154** (2010):3, 431–508.

[Gur] N. Gurevich, "The twisted Satake isomorphism and Casselman–Shalika formula", Automorphic Representation and Related Topics, conference proceedings, RIMS, Kyoto, Japan, (2013).

[GK] N. Gurevich and E. Karasiewicz,"The twisted Satake transform and the Casselman–Shalika formula for quasi-split groups", IMRN, **14** (2022), 11148–11179.

[Hai] T. Haines, "Dualities for root systems with automorphisms and applications to non-split groups", *Represent. Theory* **22** (2018), 1–26.

[HKP] T. Haines, R. Kottwiz, A. Prasad, "Iwahori–Hecke Algebras", *J. Ramanujan Math. Soc.* **25**:2 (2010), 113–145.

[HC] Harish-Chandra, "Representations of semisimple Lie groups on a Banach space", *Proc. Nat. Acad. Sci. U.S.A.* **37** (1951), 170–173.

[Ha] M. Harris, "Eisenstein series on Shimura varieties", *Ann. of Math. (2)* **119**:1 (1984), 59–94.

[HT] M. Harris and R. Taylor, *The Geometry and Cohomology of Some Simple Shimura Varieties*, Annals of Math. Studies **151** (Princeton University Press, Princeton, 2001).

[HSS] T. Hasegawa, H. Saigo, S. Saito, "A quantum probabilistic approach to Hecke algebras for p-adic PGL_2", *Infinite Dimensional Analysis, Quantum Probability and Related Topics* **21**:3 (2018).

[Hé] A. Hébert, "A new axiomatics for masures", *Canad. J. Math.* **72**:3 (2020), 732–773.

[He] E. Hecke, "Uber Modulfunktionen und die Dirichletschen Reihen mit Eulerscher Produktentwicklungen, I and II", *Math. Ann.* **114** (1937), 1–28, 316–351.

[HO] G.J. Heckman, and E.M. Opdam, "Harmonic analysis for affine Hecke algebras", in *Current Developments in Mathematics 1996*, International Press (1997), 37–60.

[Hei] V. Heiermann, "Opérateurs d'entrelacement et algèbres de Hecke avec paramètres d'un groupe réductif p-adique - le cas des groupes classiques", *Selecta Math.* **17**:3 (2011), 713–756.

[Hel1] S. Helgason, "Harish-Chandra's c-function. A mathematical jewel", in *NATO Adv. Sci. Inst. Ser. C: Math. Phys. Sci.* **429** (1994), Kluwer Acad. Publ., Dordrecht, 55–67, Reprinted in Proc. Sympos. Pure Math. **68**, Amer. Math. Soc., Providence, RI, 2000, pp. 273–283.

[Hel2] S. Helgason, *Groups and geometric analysis. Integral geometry, invariant differential operators, and spherical functions*, Corrected reprint of the 1984 original. Mathematical Surveys and Monographs, **83** Amer. Math. Soc., Providence, RI, 2000. xxii+667 pp. ISBN: 0-8218-2673-5.

[Hen] G. Henniart, "Une preuve simple des conjectures de Langlands pour $\mathrm{GL}(n)$ sur un corps p-adique", *Invent. Math.* **139** (2000, 439–455.

[HV] G. Henniart and M.-F. Vignéras, "A Satake isomorphism for representations modulo p of reductive groups over local fields", *J. Reine Angew. Math.* **701** (2015), 33–75.

[Her] F. Herzig, "A Satake isomorphism in characteristic p", *Compos. Math.* **147** (2011), 263–283.

[HII] K. Hiraga, A. Ichino, and T. Ikeda, "Formal degrees and adjoint γ-factors", *J. Amer. Math. Soc.* **21** (2008):1, 283–304, Correction: **21** (2008):4, 1211–1213.

[HS] K. Hiraga and H. Saito, "On L-packets for inner forms of SL_n", *Mem. Amer. Math. Soc.* **1013**:215 (2012).

[Hi1] Y. Hironaka, "Spherical functions of hermitian and symmetric forms I", *Japan. J. Math.* **14** (1988), 203–223.

[Hi2] Y. Hironaka, "Spherical functions of hermitian and symmetric forms II," *Japan. J. Math.* **5** (1989), 15–51.

[Hi3] Y. Hironaka, "Spherical functions of hermitian and symmetric forms III", *Tôhoku Math. J.* **40** (1988), 651–671.

[Hi4] Y. Hironaka, "Spherical functions and local densities on Hermitian forms", *J. Math. Soc. Japan* **51**:3 (1999), 553–581.

[Hi5] Y. Hironaka, "Spherical functions on p-adic homogeneous spaces.A lgebraic and analytic aspects of zeta functions and L-functions", 50–72, MSJ Mem. **21**, Mathematical Society of Japan, Tokyo, 2010.

[Hi6] Y. Hironaka, "Spherical functions and local densities on the space of p-adic quaternion Hermitian forms", *Int. J. Number Theory* **18**:3 (2022), 575–612.

[HY] Y. Hironaka and Y. Komori, "Spherical functions on the space of p-adic unitary Hermitian matrices", *Int. J. Number Theory* **10**:2 (2014), 513–558.

[HS1] Y. Hironaka and F. Satô, "Spherical functions and local densities of alternating forms", *Amer. J. Math.* **110** (1988), 473–512.

[HS2] Y. Hironaka and F. Satô, "Local densities of alternating forms", *J. Number Theory* **33** (1989), 32–52.

[HuTa] J.-S. Huang and M. Tadić, "Generalized spherical functions on reductive p-adic groups", *Trans. Amer. Math. Soc.* **357**:5 (2005), 2081–2117.

[Hu] J.E. Humphreys, *Reflection groups and Coxeter groups*, Cambridge University Press, Cambridge, 1990.

[Ih] Y. Ihara "On discrete subgroups of the two by two projective linear group over p-adic fields", *J. Math. Soc. Japan* **18** (1966), 219–235.

[IM] N. Iwahori and H. Matsumoto, "On some Bruhat decomposition and the structure of the Hecke ring of a p-adic Chevalley group", *Inst. Hautes Études Sci. Publ. Math.* **25** (1965), 5–48.

[JR] H. Jacquet and S. Rallis, "Uniqueness of linear periods", *Compositio Math.* **102**:1 (1996), 65–123.

[JLZ] A. Joseph, G. Letzter and S. Zelikson, "On the Brylinski-Kostant filtration", *J. Amer. Math. Soc.* **13**:4 (2000), 945—970.

[KP] T. Kaletha and G. Prasad, *Bruhat–Tits theory: a new approach*, New Math. Monographs **44**, Cambridge University Press 2023.

[Kak] H. Kakuhama, "Formal degrees and the local theta correspondence: The quaternionic case", *Represent. Theory* **26** (2022), 1192–1267.

[Kat] S. Kato, "Spherical functions and a q-analogue of Kostant's weight multiplicity formula", *Invent. Math.* **66** (1982), 461–468.

[KSM] S. Kato, A. Murase and T. Sugano, *Whittaker–Shintani functions for orthogonal groups*, *Tohoku Math. J.* **55** (2003), 1–64.

[KL1] D. Kazhdan and G. Lusztig, "Representations of Coxeter groups and Hecke algebras", *Invent. Math.* **53**:2, 165–184 (1979).

[KL2] D. Kazhdan and G. Lusztig, "Schubert varieties and Poincaré duality", in: *Geometry of the Laplace operator*, Proc. Sympos. Pure Math., vol. XXXVI, pp. 185–203. Amer. Math. Soc., Providence (1980).

[Ko] T.H. Koornwinder, "Askey–Wilson polynomials as zonal spherical functions on the SU(2) quantum group", *SIAM J. Math. Anal.* **24**:3 (1993), 795–813.

[Kot] R. Kottwitz, "Tamagawa numbers", *Ann. of Math.* **127** (1988), 629–646.

[Kn] F. Knop, "The Luna–Vust theory of spherical embeddings", in *Proc. Hyderabad Conference on Algebraic Groups, Hyderabad, 1989*, (Manoj Prakashan, Madras, 1991), 225–249.

[Laf] V. Lafforgue, "KK-théorie bivariante pour les algèbres de Banach et conjecture de Baum–Connes", *Invent. Math.* **149**:1 (2002), 1–95.

[LRS] G. Laumon, M. Rapoport, and U. Stuhler, "D-elliptic sheaves and the Langlands correspondence", *Invent. Math.* **113** (1993), 217–238.

[LRZ] Ch. Li, M. Rapoport and W. Zhang, "Arithmetic Fundamental Lemma for the spherical Hecke algebra", *Manuscripta Math.* **175** (2024), 1–51..

[Lus1] G. Lusztig, "Singularities, character formulas, and a q-analog of weight multiplicities", in *Analysis and Topology on Singular Spaces. II, III* (Luminy, 1981), Astérisque **101**, 208–229. Soc. Math. France, Paris (1983).

[Lus2] G. Lusztig, "Affine Hecke algebras and their graded version", *J. Amer. Math. Soc.* **2** (3) (1989), 599–635.

[Lus3] G. Lusztig, "Classification of unipotent representations of simple p-adic groups", *Int. Math. Res. Notices* **11** (1995), 517–589.

[Lus4] G. Lusztig, "Classification of unipotent representations of simple p-adic groups. II", *Represent. Theory* **6** (2002), 243–289.

[Lus5] G. Lusztig, "The history and significance of the Bruhat decomposition of a reductive group", Preprint arXiv:1006.5004 (2010).

[Mac1] I.G. Macdonald, "The Poincaré series of a Coxeter group", *Math. Ann.* **199** (1972), 161–174.

[Mac2] I.G. Macdonald, *Symmetric Functions and Hall Polynomials*, 2nd Edition, Clarendon Press, Oxford, 1995.

[Mac3] I.G. Macdonald, "Orthogonal polynomials associated with root systems", *Séminaire Lotharingien de Combinatoire* **45** (2001), Article B45a.

[Mac4] I.G. Macdonald, *Affine Hecke algebras and orthogonal polynomials*, Cambridge Tracts in Math. **157**, Cambridge University Press, Cambridge 2003.

[Mar] T. Marquis, *An introduction to Kac–Moody groups over fields*, EMS Textbooks in Mathematics, European Mathematical Society, Zurich (2018).

[Mau1] F.I. Mautner, "Fourier analysis and symmetric spaces", *Proc. Nat. Acad. Sci. U.S.A.* **37** (1951), 529–533.

[Mau2] F.I. Mautner, "Spherical functions over p-adic fields, I", In: *Amer. J. Math.* **80**:2 (1958), 441–457.

[Mau3] F.I. Mautner, "Spherical functions over p-adic fields, II", In: *Amer. J. Math.* **86**:1 (1964), 171–200.

[May] A. Mayeux,"Bruhat–Tits theory from Berkovich's point of view. Analytic filtrations", *Ann. H. Lebesgue* **5** (2022), 813–839.

[Miz] H. Mizukawa, "Zonal spherical functions on the complex reflection groups and $(n+1, m+1)$-hypergeometric functions", *Adv. Math.* **184** (2004), 1–17.

[MP1] A. Moy and G. Prasad, "Unrefined minimal K-types for p-adic groups", *Invent. Math.* **116** (1994), 393–408.

[MP2] A. Moy and G. Prasad,"Jacquet functors and unrefined minimal K-types", *Comment. Math. Helv.* **71**:1 (1996), 98–121.

[MV] I. Mirković and K. Vilonen, "Geometric Langlands duality and representations of algebraic groups over commutative rings", *Ann. Math. (2)* **166** (1), 95–143 (2007).

[NR] K. Nelsen and A. Ram, "Kostka–Foulkes polynomials and Macdonald spherical functions", in *Surveys in combinatorics*, London Math. Soc. Lecture Note Ser., **307**, 325–370. Cambridge Univ. Press, Cambridge, 2003.

[N] Bao Châu Ngô, "Le lemme fondamental pour les algèbres de Lie", *Publ. Math. Inst. Hautes Etudes Sci.* **111** (2010), 1–169.

[Off] O. Offen, "Relative spherical functions on p-adic symmetric spaces (three cases)", *Pacific J. Math.* **215**:1 (2004), 97–149. *Erratum*: *Pacific J. Math.* **236**:1 (2008), 195–200.

[Oh] K. Ohara, "On the formal degree conjecture for non-singular supercuspidal representations", *IMRN*, Issue 13 (2023), 10997–11034.

[OST] M. Oi, R. Sakamoto, and H. Tamori, "Iwahori–Hecke algebra and unramified local L-functions", *Math. Zeitschrift* **30**:3 (2023), p. 59.

[Op1] E. Opdam, "Harmonic analysis for certain representations of graded Hecke algebras", *Acta Math.* **175** (1995), 75–121.

[Op2] E. Opdam, *Lecture notes on Dunkl operators for real and complex reflection groups*, MSJ Memoirs **8**, Mathematical Society of Japan, Tokyo, 2000.

[Op3] E. Opdam, "A generating formula for the trace of the Iwahori–Hecke algebra", *Progr. Math.* **210** (2003), 301–323.

[Op4] E. Opdam, "On the spectral decomposition of affine Hecke algebras", *J. Inst. Math. Jussieu* **3**:4 (2004), 531–648.

[Op5] E. Opdam, "Hecke algebras and harmonic analysis", in International Congress of Mathematicians. Vol. II, pp. 1227–1259. Eur. Math. Soc. (EMS), Zürich, 2006.

[Op6] E. Opdam, "Affine Hecke algebras and the conjectures of Hiraga, Ichino and Ikeda on the Plancherel density", Conference in honor of Joseph Berstein, Proc. Sympos. Pure Math. **101**, 309–350, Amer. Math. Soc. 2019.

[Par1] J. Parkinson, "Buildings and Hecke algebras", *Journal of Algebra* **297** (2006), 1–49.

[Par2] J. Parkinson, "Spherical harmonic analysis on affine buildings", *Math. Z.* **253** (2006), 571–606.

[Par3] J. Parkinson, "Isotropic random walks on affine buildings", *Ann. Inst. Fourier* **57** (2007), 379–419.

[Pet] C. Peterson, *Quantum ergodicity on Bruhat–Tits buildings*, Ph.D thesis University of Michigan, 2023.

[PY] G. Prasad and J.-K. Yu, "On quasi-reductive group schemes", *J. Algebraic Geom.* **15**:3 (2006), 507–549, with an appendix by Brian Conrad.

[Ra] A. Ram, "Alcove walks, Hecke Algebras, spherical functions, crystals and column strict tableaux", *Pure and Applied Mathematics Quarterly* **2**:4 (Special Issue: In honor of Robert MacPherson), 135–183, 2006.

[Ré] B. Rémy, "Groupes de Kac–Moody déployés et presque déployés", Astérisque (2002), **277**, viii+348 pp.

[RTW] B. Rémy, A. Thuillier, and A. Werner, "Bruhat–Tits theory from Berkovich's point of view. I. Realizations and compactifications of buildings", *Ann. Sci. Éc. Norm. Supér. (4)*, **43** (2010), 461–554.

[Ren] D. Renard, *Représentations des groupes réductifs p-adiques*, Cours Spécialisés **17**, Société Mathématique de France, Paris, 2010. vi+332 pp. ISBN: 978-285629-278-5.

[Ri] T. Richarz, "A new approach to the geometric Satake equivalence", *Doc. Math.* **19** (2014), 209–246.

[Ro1] A. Roche, "Parabolic induction and the Bernstein decomposition", *Compositio Math.* **134** :2(2002), 113–133.

[Ro2] A. Roche, "The Bernstein decomposition and the Bernstein centre", Ottawa lectures on admissible representations of reductive p-adic groups, Fields Inst. Monogr., vol. **26**, Amer. Math. Soc., Providence, RI, 2009, 3–52.

[Ron] M. Ronan, *Lectures on buildings: Updated and Revised*, Perspectives in Mathematics **7**, Academic Press, 2009.

[Sa1] Y. Sakellaridis, "A Casselman–Shalika formula for the Shalika model of GL_n", *Canad. J. Math.* **58**:5 (2006), 1095–1120.

[Sa2] Y. Sakellaridis,"On the unramified spectrum of spherical varieties over p-adic fields", *Compositio Math.* **144**:4 (2008), 978–1016.

[Sa3] Y. Sakellaridis, "Spherical functions on spherical varieties", *Amer. J. Math.* **135**:5 (2013), 1291–1381.

[SV] Y. Sakellaridis and A. Venkatesh, "Periods and harmonic analysis on spherical varieties", Astérisque **396** (2017), viii+360 pp.

[SS] P. Schneider and U. Stuhler, "Representation theory and sheaves on the Bruhat–Tits building", *Publ. math. I.H.É.S.* **85** (1997), 97–191.

[Scho] P. Scholze,"The local Langlands correspondence for GL_n over p-adic fields", *Invent. Math.* **192** (2013), 663–715.

[SGA3] M. Artin, J.E. Bertin, M. Demazure, P. Gabriel, A. Grothendieck, M. Raynaud, and J.-P. Serre, Schémas en Groupes. I, II, III, Lecture Notes in Mathematics, **151**, **152**, **153**, Springer-Verlag, New York, 1970.

[Sha1] F. Shahidi, "On certain L-functions", *Amer. J. Math.* **103** (1981), 297–355.

[Sha2] F. Shahidi, "A proof of Langlands' conjecture on Plancherel measures; complementary series for p-adic groups", *Ann. of Math. (2)* **132**:2 (1990), 273–330.

[Sil] A.J. Silberger, *Introduction to harmonic analysis on reductive p-adic groups*, Mathematical Notes, vol. **23**, Princeton University Press, Princeton, N.J.; University of Tokyo Press, Tokyo, 1979.

[So1] M. Solleveld, "A local Langlands correspondence for unipotent representations", *Amer. J. Math.* **145**:3 (2023), 673–719.

[So2] M. Solleveld, "Endomorphism algebras and Hecke algebras for reductive p-adic groups", *J. Algebra* **606** (2022), 371–470.

[Tad] M. Tadić, "Harmonic analysis of spherical functions on reductive groups over p-adic fields", *Pacific J. Math.* 1983.

[Tat] J. Tate, "Number theoretic background", Proceedings of Symposia in Pure Mathematics **33** (1979), part 2, 3–26.

[Ti1] J. Tits, *Buildings of spherical type and finite BN-pairs*, Lecture Notes in Mathematics **386**, Springer-Verlag, 1974.

[Ti2] J. Tits, "Reductive groups over local fields", in *Automorphic Forms, Representations and L-Functions* (Proc. Sympos. Pure Math., Oregon State Univ., Corvallis, Ore., 1977), Part 1, Proc. Sympos. Pure Math., XXXIII, Amer. Math. Soc., Providence, R.I., 1979, pp. 29–69.

[T3] J. Tits, "Immeubles de type affine", in: Buildings and the Geometry of Diagrams (Como1984), ed. by L.A. Rosati, Lecture Notes in Math. **1181**, Springer-Verlag, Berlin 1986, 159–190.

[Ti4] J. Tits, "Uniqueness and presentation of Kac–Moody groups over fields", *J. of Algebra* **105** (1987), 542–573.

[Tr] B. Trojan, "Asymptotic behavior of heat kernels and Green functions on affine buildings", *J. Eur. Math. Soc.* (2024).

[vDi] G. Van Dijk, "Harmonic analysis on reductive p-adic groups (after Harish-Chandra)", Séminaire Bourbaki, 1970–71.

[Vin] B. Vinberg, "Complexity of actions of reductive groups", *Funktsional. Anal. i Prilozhen.* **20** (1986), 1–13, 96.

[Wal] J.-L. Waldspurger, "La formule de Plancherel pour les groupes p-adiques (d'après Harish-Chandra)", *J. Inst. Math. Jussieu* **2**:2 (2003), 235–333.

[Wei1] R. Weiss, *The structure of spherical buildings*, Princeton University Press, 2003.

[Wei2] R. Weiss, *The structure of affine buildings*, Princeton University Press, 2008.

[Zha] W. Zhang, "On arithmetic fundamental lemmas", *Invent. Math.* **188**:1 (2012), 197–252.

[Zhu] X. Zhu, *An introduction to affine Grassmannians and the geometric Satake equivalence*, IAS/Park City Math. Ser. **24**, Amer. Math. Soci. Providence, RI, 2017, 59–154.

Index

I. G. Macdonald and A.-M. Aubert, *Spherical Functions on a Group of p-adic Type*,
Lecture Notes in Mathematics 2392,
https://doi.org/10.1007/978-3-032-15671-6

GPSR Compliance

The European Union's (EU) General Product Safety Regulation (GPSR) is a set of rules that requires consumer products to be safe and our obligations to ensure this.

If you have any concerns about our products, you can contact us on ProductSafety@springernature.com

In case Publisher is established outside the EU, the EU authorized representative is:

Springer Nature Customer Service Center GmbH
Europaplatz 3
69115 Heidelberg, Germany

Batch number: 10370708

Printed by Printforce, the Netherlands